Analysis of Sensory Properties in Foods

Analysis of Sensory Properties in Foods

Special Issue Editor
Edgar Chambers IV

MDPI • Basel • Beijing • Wuhan • Barcelona • Belgrade

Special Issue Editor
Edgar Chambers IV
Kansas State University
USA

Editorial Office
MDPI
St. Alban-Anlage 66
4052 Basel, Switzerland

This is a reprint of articles from the Special Issue published online in the open access journal *Foods* (ISSN 2304-8158) from 2018 to 2019 (available at: https://www.mdpi.com/journal/foods/special_issues/Sensory_Analysis_Foods).

For citation purposes, cite each article independently as indicated on the article page online and as indicated below:

LastName, A.A.; LastName, B.B.; LastName, C.C. Article Title. *Journal Name* **Year**, *Article Number*, Page Range.

ISBN 978-3-03921-433-4 (Pbk)
ISBN 978-3-03921-434-1 (PDF)

© 2019 by the authors. Articles in this book are Open Access and distributed under the Creative Commons Attribution (CC BY) license, which allows users to download, copy and build upon published articles, as long as the author and publisher are properly credited, which ensures maximum dissemination and a wider impact of our publications.

The book as a whole is distributed by MDPI under the terms and conditions of the Creative Commons license CC BY-NC-ND.

Contents

About the Special Issue Editor vii

Edgar Chambers IV
Analysis of Sensory Properties in Foods: A Special Issue
Reprinted from: *foods* **2019**, *8*, 291, doi:10.3390/foods8080291 1

Suntaree Suwonsichon
The Importance of Sensory Lexicons for Research and Development of Food Products
Reprinted from: *foods* **2019**, *8*, 27, doi:10.3390/foods8010027 4

Carla Kuesten and Jian Bi
Temporal Drivers of Liking Based on Functional Data Analysis and Non-Additive Models for
Multi- Attribute Time-Intensity Data of Fruit Chews
Reprinted from: *foods* **2018**, *7*, 84, doi:10.3390/foods7060084 20

Damir Dennis Torrico, Wannita Jirangrat, Jing Wang, Penkwan Chompreeda, Sujinda Sriwattana and Witoon Prinyawiwatkul
Novel Modelling Approaches to Characterize and Quantify Carryover Effects on
Sensory Acceptability
Reprinted from: *foods* **2018**, *7*, 186, doi:10.3390/foods7110186 37

Julia Dirler, Gertrud Winkler and Dirk W. Lachenmeier
What Temperature of Coffee Exceeds the Pain Threshold? Pilot Study of a Sensory Analysis
Method as Basis for Cancer Risk Assessment
Reprinted from: *foods* **2018**, *7*, 83, doi:10.3390/foods7060083 48

Jiyun Yang and Jeehyun Lee
Application of Sensory Descriptive Analysis and Consumer Studies to Investigate Traditional
and Authentic Foods: A Review
Reprinted from: *foods* **2019**, *8*, 54, doi:10.3390/foods8020054 60

Shangci Wang, Shaokang Zhang and Koushik Adhikari
Influence of Monosodium Glutamate and Its Substitutes on Sensory Characteristics and
Consumer Perceptions of Chicken Soup
Reprinted from: *foods* **2019**, *8*, 71, doi:10.3390/foods8020071 77

Viktoria Olsson, Andreas Håkansson, Jeanette Purhagen and Karin Wendin
The Effect of Emulsion Intensity on Selected Sensory and Instrumental Texture Properties of
Full-Fat Mayonnaise
Reprinted from: *foods* **2018**, *7*, 9, doi:10.3390/foods7010009 93

Leontina Lipan, Marina Cano-Lamadrid, Mireia Corell, Esther Sendra, Francisca Hernández, Laura Stan, Dan Cristian Vodnar, Laura Vázquez-Araújo and Ángel A. Carbonell-Barrachina
Sensory Profile and Acceptability of HydroSOStainable Almonds
Reprinted from: *foods* **2019**, *8*, 64, doi:10.3390/foods8020064 102

Edgar Chambers V, Edgar Chambers IV and Mauricio Castro
What Is "Natural"? Consumer Responses to Selected Ingredients
Reprinted from: *foods* **2018**, *7*, 65, doi:10.3390/foods7040065 116

About the Special Issue Editor

Edgar Chambers IV is Distinguished Professor of Sensory Analysis and Consumer Behavior at Kansas State University. Previously, he was in industry, including time as Manager of Sensory and Statistical Analyses at the world-headquarters of the Seven-Up Company. He consults extensively with industry, government, and nonprofits worldwide on product evaluation and consumer understanding, teaches graduate classes in sensory analysis and consumer behavior and directs research projects in product evaluation and consumer understanding. His expertise encompasses food products such as meat and grains, packaging, personal care, fabric, paper, pharmaceutical, paint, and other consumer and industrial products. Dr. Chambers is past Chair of both the Society of Sensory Professionals and the Sensory Division of ASTM International and was named a "Fellow" of ASTM in 2006. He received the David Peryam Award, the highest award in his field, in 2006. He is the Editor-in-Chief of Beverages and the former editor of Journal of Sensory Studies.

Editorial

Analysis of Sensory Properties in Foods: A Special Issue

Edgar Chambers IV

Center for Sensory Analysis and Consumer Behavior, Kansas State University, 1310 Research Park Dr., Manhattan, KS 66502, USA; eciv@ksu.edu

Received: 24 July 2019; Accepted: 25 July 2019; Published: 26 July 2019

Abstract: The sensory properties of foods are the most important reason people eat the foods they eat. What those properties are and how we best measure those properties are critical to understanding food and eating behavior. Appearance, flavor, texture, and even the sounds of food can impart a desire to eat or cause us to dismiss the food as unappetizing, stale, or even inappropriate from a cultural standpoint. This special issue focuses on how sensory properties, including consumer perceptions, are measured, the specific sensory properties of various foods, which properties might be most important in certain situations, and how consumers use sensory attributes and consumer information to make decisions about what they believe about food and what they will eat.

Keywords: sensory; foods; consumer; descriptive

1. Introduction

Sensory analysis is an interdisciplinary science comprised of information and methods adapted from psychology, physiology, statistics, linguistics, food science, nutrition, medicine, chemistry, physics, sociology, anthropology, and a host of other fields. The antecedents of sensory testing go back many millennia, but modern testing of sensory properties of foods really began in earnest after World War I when the United States (US) military realized that soldiers came back from combat malnourished. This was caused, in part, because the food that was available to soldiers in military kitchens and through military rations had such poor sensory quality that the soldiers refused to eat it. In 1953 a symposium held in Chicago by the US Quartermaster Food and Container Institute of the armed forces was held to bring together various groups working to conduct sensory (including consumer) testing of foods [1]. In the proceedings of that conference, the organizers state "the impact of food testing methods has been felt across the nation. … the quality of food served both to the civilian population and to the Armed Services has been improved. Pretesting of new items and quality control testing of established products have already provided consumers with a more uniformly excellent food quality." [1]

In the 1940s the US Army quartermaster corps scientists began studying human acceptance and how to measure it [2]. At the same time, scientists at Arthur D. Little, Inc. began promoting the use of descriptive sensory methods [3] for quantitatively measuring the sensory perception of food attributes. Cover [4] had already published work on a discrimination test, now called the paired comparison. In 1970 Mina McDaniel started work [5] on what could arguably be called the first dissertation in sensory sciences from within the food science field at the University of Massachusetts, Amherst. My own doctoral dissertation in 1980 and the resulting publication [6] was a short tome of just 36 pages with fewer than 20 references. There were few references because so little real science had been conducted at that point on sensory methods related to testing of actual foods. Heymann [7] recently published a "history" of sensory focusing primarily on the time since the 1940s and the many advancements made in sensory analysis since then. She comments on the work of various pioneers in the field. She mentions scientific organizations have taken hold, major conferences are being held, and scientists around the world are focusing on issues of product attributes, consumer acceptance and behavior, and ways in which to measure those aspects with more accurate and meaningful data. Yet, with the thousands of papers published in all those years, we still harken back to such fundamental papers as those describing the hedonic scale [2] and the flavor profile [8]. Fundamental information is still needed on many things.

Much of the sensory information we need can be divided into three categories, all of which are touched on in this group of papers. First, are studies that impact sensory methods. In science, methods are critical; without good methods we cannot collect good data. Four papers in this collection focus primarily on methods. Suwonsichon [9] treats us to a recent review of descriptive attributes. She updates older literature with papers in the past 5 years that focus on the description of various products including both human and pet foods. Such studies of attributes are essential for scientists to collect accurate, reproducible information on products across laboratories and countries. Of course, not all attributes happen simultaneously and Kuesten and Bi [10] provide us with ways to analyze data for multiattribute time intensity, particularly when used to compare to consumer acceptance in studies of the things that drive liking and disliking. For product developers, such information is critical in order to understand the attributes that must be present and at what levels to increase liking as well as to know which attributes should be reduced or eliminated. Torrico and others [11] provide us with information on carryover effects that can result in data that is less accurate because of sampling of prior products. Much has been written on the effects of this type of bias in all types of sensory studies, but this paper focuses on statistical ways to determine whether it may have occurred and for which attributes in consumer studies. Lastly, the temperature at which pain is associated with drinking coffee [12] provides both a practical method for assessing such perceptions as well as practical information on consumer liking of temperature when drinking coffee. The authors show that many consumers like coffee to be around 63 °C at consumption, which is only slightly less than the temperature (67 °C) that brings pain to many consumers. Furthermore, they tell us that, unfortunately, those temperatures are similar to the temperature at which hot foods/beverages can result in carcinogenicity.

The second area of focus is on the evaluation of products either by trained panels or consumers. This section is led by Yang and Lee's review [13] of the evaluation of traditional and authentic foods. Those authors highlight a number of studies of products that are considered "ethnic" in nature, such as kimchi, artisanal cheeses, traditional sausages, and many more. Those types of products are key both within the countries and cultures they traditionally are consumed, but because food becomes more "global" and is introduced into new countries and cultures it must remain authentic, yet be appropriate for the new consumer. Wang and others focus on the effects of monosodium glutamate (MSG) in foods and the impact on flavor characteristics as well as liking [14]. Such information is helpful in understanding the nature of particular ingredients and their effect on both specific sensory properties as well as consumer acceptance. A further paper by Olsson et al. [15] shows the impact of processing on a specific food, in this case, the impact of emulsification intensity on sensory properties of mayonnaise. That paper is an example of the impact that processing can have on a product's flavor and texture. The effect of water sustainability when growing almonds is evaluated by Lipan et al. [16]. This paper shows that water use can be reduced when growing almonds without an impact on the sensory properties or consumer liking of the nuts. The paper also transcends the description of the product and moves us into the next type of paper covered in this special issue, that of consumer behavior. The authors measured consumer acceptance of the concept of HydoSOStainable almonds and showed that consumers were potentially willing to pay more for the product. This is good news for growers and processors who could make more money with a product that, even though it is sensorially the same, has benefits that appeal to consumers outside the sensory aspects.

In addition to the research on sustainable almonds, another paper in this issue covers perception as it relates to behavior. The final paper in the issue focuses on "naturalness" of food ingredients [17]. Recent consumer trends have embraced the concepts of sustainability, organic production, and healthfulness.

Although many people consider these aspects to make products "natural" or at least boost "naturalness", the authors of this paper show clear evidence that consumers make assumptions about naturalness based on ingredients they may not understand. The authors point out that not understanding what an ingredient is, the use of "chemical-sounding names" and other issues impact whether consumers consider an ingredient natural, regardless of its actual source and processing.

These papers reference literally hundreds of other papers containing sensory data, which is an enormous leap from even 40 years ago. The world of sensory analysis continues to make headway into helping maintain a food supply that not only nourishes our bodies but also satisfies our minds and brings pleasure to our lives.

Funding: This research received no external funding.

Conflicts of Interest: The author declares no conflict of interest.

References

1. Peryam, D.R.; Pilgrim, F.J.; Peterson, M.S. (Eds.) *Food Acceptance Testing Methodology*; National Academy of Sciences, National Research Council: Washington, DC, USA, 1953.
2. Peryam, D.R.; Pilgrim, F.J. Hedonic scale method of measuring food preferences. *Food Technol.* **1957**, *11*, 9–14.
3. Cairncross, S.E.; Sjöström, L.B. Flavor profiles—A new approach to flavor problems. *Food Technol.* **1950**, *4*, 308–311.
4. Cover, S. A new subjective measurement testing tenderness in meat—The paried eating method. *J. Food Sci.* **1936**, *1*, 287–295. [CrossRef]
5. Oregon State University Oral History Project. Mina McDaniel Oral History Interview. Available online: http://scarc.library.oregonstate.edu/oh150/mcdaniel/biography.html. (accessed on 23 July 2019).
6. Chambers, E., IV; Bowers, J.A.; Dayton, A.D. Statistical designs and panel training/experience for sensory analysis. *J. Food Sci.* **1981**, *46*, 1902–1906. [CrossRef]
7. Heymann, H. A personal history of sensory science. *Food Cult. Soc.* **2019**, *22*, 203–223. [CrossRef]
8. Caul, J.F. The profile method of flavor analysis. In *Advances in Food Research*; Mrak, E.M., Stewart, G.F., Eds.; Academic Press: New York, NY, USA, 1957; pp. 1–40.
9. Suwonsichon, S. The Importance of Sensory Lexicons for Research and Development of Food Products. *Foods* **2019**, *8*, 27. [CrossRef] [PubMed]
10. Kuesten, C.; Bi, J. Temporal Drivers of Liking Based on Functional Data Analysis and Non-Additive Models for Multi-Attribute Time-Intensity Data of Fruit Chews. *Foods* **2018**, *7*, 84. [CrossRef] [PubMed]
11. Torrico, D.D.; Jirangrat, W.; Wang, J.; Chompreeda, P.; Sriwattana, S.; Prinyawiwatkul, W. Novel Modelling Approaches to Characterize and Quantify Carryover Effects on Sensory Acceptability. *Foods* **2018**, *7*, 186. [CrossRef] [PubMed]
12. Dirler, J.; Winkler, G.; Lachenmeier, D.W. What Temperature of Coffee Exceeds the Pain Threshold? Pilot Study of a Sensory Analysis Method as Basis for Cancer Risk Assessment. *Foods* **2018**, *7*, 83. [CrossRef] [PubMed]
13. Yang, J.; Lee, J. Application of Sensory Descriptive Analysis and Consumer Studies to Investigate Traditional and Authentic Foods: A Review. *Foods* **2019**, *8*, 54. [CrossRef] [PubMed]
14. Wang, S.; Zhang, S.; Adhikari, K. Influence of Monosodium Glutamate and Its Substitutes on Sensory Characteristics and Consumer Perceptions of Chicken Soup. *Foods* **2019**, *8*, 71. [CrossRef] [PubMed]
15. Olsson, V.; Håkansson, A.; Purhagen, J.; Wendin, K. The Effect of Emulsion Intensity on Selected Sensory and Instrumental Texture Properties of Full-Fat Mayonnaise. *Foods* **2018**, *7*, 9. [CrossRef] [PubMed]
16. Lipan, L.; Cano-Lamadrid, M.; Corell, M.; Sendra, E.; Hernández, F.; Stan, L.; Vodnar, D.C.; Vázquez-Araújo, L.; Carbonell-Barrachina, Á.A. Sensory Profile and Acceptability of HydroSOStainable Almonds. *Foods* **2019**, *8*, 64. [CrossRef] [PubMed]
17. Chambers, E., V; Chambers, E., IV; Castro, M. What Is "Natural"? Consumer Responses to Selected Ingredients. *Foods* **2018**, *7*, 65. [CrossRef] [PubMed]

© 2019 by the author. Licensee MDPI, Basel, Switzerland. This article is an open access article distributed under the terms and conditions of the Creative Commons Attribution (CC BY) license (http://creativecommons.org/licenses/by/4.0/).

Review

The Importance of Sensory Lexicons for Research and Development of Food Products

Suntaree Suwonsichon *

Kasetsart University Sensory and Consumer Research Center, Department of Product Development, Faculty of Agro-Industry, Kasetsart University, Bangkok 10900, Thailand

Received: 28 December 2018; Accepted: 10 January 2019; Published: 15 January 2019

Abstract: A lexicon is a set of standardized vocabularies developed by highly trained panelists for describing a wide array of sensory attributes present in a product. A number of lexicons have been developed to document and describe sensory perception of a variety of food categories. The current review provides examples of recently developed sensory lexicons for fruits and vegetables; grains and nuts; beverages; bakery, dairy, soy and meat products; and foods for animals. Applications of sensory lexicons as an effective communication tool and a guidance tool for new product development processes, quality control, product improvement, measuring changes during product shelf life, and breeding new plant cultivars are also discussed and demonstrated through research in the field.

Keywords: sensory; descriptive analysis; lexicon; food; product development; shelf life; quality control; product improvement; plant breeding

1. Introduction

Descriptive sensory analysis is the most powerful method for capturing a product's characteristics in terms of their perceived attributes and intensities. This method has been used extensively to characterize the sensory characteristics of various food products [1]. A lexicon is a set of standardized vocabularies developed and used by panelists who are highly trained for describing a wide array of sensory attributes present in a product. Lexicon development is one of the crucial steps in descriptive sensory analysis. To develop a lexicon, the panelists evaluate samples that, as much as possible, represent the entire product space, generate terms that describe the samples, define the terms, develop standardized evaluation procedures, select references that clarify the terms, review samples to further train the panelists, and then finalize those terms [2]. In most cases, panelists also assign a score for each reference standard to anchor an attribute scale. Terms listed in the lexicon must be extensive and complete, non-hedonic, singular (not integrated), and non-redundant, and also must capture all product differences [3]. Previously, Lawless and Civille [2] and Drake and Civille [4] outlined important considerations for lexicon development and reviewed a number of food lexicons published before the year 2013.

A sensory lexicon plays an important role as an effective communication tool among diverse audiences such as sensory panels, sensory scientists, product developers, marketing professionals, and suppliers who may have different understandings of the same sensory attribute due to differences in perception, background knowledge, and culture. In addition, carrying out descriptive sensory analysis using a lexicon with well-defined and referenced descriptors and standardized evaluation procedures provide accurate and repeatable information about sensory qualities of food products that could be used as a guidance tool for various activities in the research and development of food products, including new product development, quality control, product improvement, measuring changes during product shelf-life, and breeding new plant cultivars.

The aims of this article were to: (1) review recently developed food lexicons, and (2) discuss and exemplify how the lexicons could be used to strengthen research and development of food products.

2. Recent Lexicons for Foods

A number of lexicons have been developed to describe sensory characteristics of a variety of food products, including: fruits and vegetables; grains and nuts; beverages; bakery, dairy, soy and meat products; and foods for animals. Some lexicons further explored complex sensory attributes (e.g., smoky) that may have a different character depending on the food product. Examples of recent lexicons of various food categories published mostly during the year 2013–2018 are as follows. All of them were developed by trained panelists using a descriptive sensory analysis approach. Lexicons may be developed as part of the process to evaluate a set of products, and may make that lexicon particularly useful for other researcher's projects if the tested product set was large and encompassed a wide range of samples. Lexicons may also be developed as a stand-alone project in an attempt to provide a starting point for many researchers to have a common ground on which to build studies that can be compared. Both types of lexicon are included in this review. Papers that included terminology or groups of words developed by consumers, minimally trained panelists, or that contained only a few terms or included liking or quality characteristics were not included in this review.

2.1. Fruits and Vegetables

Recent examples of lexicons for fruits, vegetables, and related products include those for apple, pomelo, peach, blackberry, strawberry, pomegranate, mango puree, sweet tamarind, and cabbage kimchi. Corollaro et al. [5] defined and referenced 15 terms to describe sensory differences, mainly texture properties, among apple cultivars and perceivable changes of the fruit during refrigerated storage. More recently, Bowen et al. [6] examined 78 apple cultivars and developed a lexicon for describing the diversity in flavor and texture characteristics of the fruit among cultivars. For pomelo, 30 descriptors were developed to distinguish between fruit of different cultivars at various fresh-cut storage periods [7]. Belisle et al. [8] established a lexicon containing aroma, flavor, and texture attributes that described differences among cultivars and stages of ripeness of peach (mature, under-ripe, over-ripe). Du et al. [9] developed a lexicon for describing aroma characteristics of blackberry of seven commercial cultivars and two newly-bred cultivars. Oliver et al. [10] defined and referenced 25 terms for characterizing sensory properties of six Australian strawberry cultivars at two maturation stages (under-ripe and ideal maturation stage). Vázquez-Araújo et al. [11] developed 35 descriptors with definitions and references for describing flavor and texture characteristics of 20 pomegranate cultivars. A lexicon was developed for comparison of flavor and texture properties between fresh mangoes and heat-treated mango purées of six mango cultivars [12]. A lexicon for sweet tamarind, a major edible fruit and flavoring ingredient in Asia, was established based on six sweet tamarind cultivars [13]. The lexicon revealed 14 attributes that were present in all cultivars and seven attributes that provided uniqueness for some cultivars. Chambers et al. [14] developed a lexicon of 17 attributes that distinguished flavor and key texture characteristics of cabbage (*baechu*) kimchi (traditional Korean side dish) of wedge vs. sliced types and fresh vs. fermented types. The attribute terms were given in English and Korean. A lexicon for dried fig was developed by Haug et al. [15] and it contained 68 terms for describing interior and exterior appearance, aroma, flavor, and aftertaste of dried fig samples of different cultivars.

2.2. Grains and Nuts

Miller and Chambers [16] examined seven black walnut cultivars and developed a lexicon that described flavor differences among cultivars. Lynch et al. [17] further expanded the lexicon of Miller and Chambers [16] by examining three additional black walnut cultivars. An additional flavor attribute (banana-like) was added to the lexicon, along with terms for describing appearance, aroma, and texture characteristics of black walnut samples. The lexicon for pecan included 20 terms with which to describe flavor differences between cultivars in raw and roasted forms [18]. Griffin et al. [19] established a lexicon containing 29 attributes that allowed for characterization of flavor and texture differences among various cashew-nut samples, including raw, oil-roasted, dry-roasted, skin-on, and rancid types. A lexicon for quinoa, a pseudo-cereal similar to amaranth and buckwheat, was developed using 21 commercial quinoa varieties, and it consisted of 27 terms for describing variations in color, aroma, flavor, and texture characteristics among quinoa varieties [20]. More recently, a lexicon for cooked spaghetti was developed using a large sample set of spaghetti from various countries. The lexicon listed 35 attributes, of which 19 were for texture; this indicated a wide variation in texture characteristics among cooked spaghetti samples [21].

2.3. Beverages

A lexicon consisting of 32 aroma/flavor, taste, and mouthfeel attributes was developed for freshly pressed and processed blueberry juice [22]. Kim et al. [23] identified, defined, and referenced 23 terms that described the flavor and mouthfeel of commercial orange juice products sold in Korea. Bhumiratana et al. [24] developed a lexicon consisting of 15 terms for describing aroma characteristics of coffee samples at each preparation step (green coffee bean, roasted coffee bean, ground coffee, and brewed coffee) as affected by coffee varieties and roasting levels. Sanchez et al. [25] identified, defined, and referenced 28 terms for describing the aroma, flavor, and aftertaste of brewed coffee of different coffee varieties and brewing methods. More recently, Chambers et al. [26] developed a lexicon containing a large set of 110 terms for describing the aroma and flavor of a wide range of brewed coffee samples. Their coffee lexicon was created based on the evaluation of 105 coffee samples from 14 countries around the world. The researchers also organized all attributes into a coffee tree for a better understanding of coffee characteristics. For hibiscus tea, samples including freshly prepared and ready-to-drink-infusions, syrups, concentrates, and instant tea were evaluated, and 21 descriptors were defined and referenced to describe the appearance, aroma/flavor, and mouthfeel of the samples [27]. All descriptors were also assembled into a sensory wheel. A lexicon for pink port wine was developed by Monteiro et al. [28] and it consisted of 21 descriptors with which to describe differences in appearance, aroma, flavor, mouthfeel, and aftertaste of the wine samples from five different brands.

2.4. Bakery Products

Morais et al. [29] established a lexicon containing 15 terms for describing variations in the appearance, aroma, flavor, and texture of pre-biotic gluten-free bread formulated with different types and concentrations of sweeteners and prebiotics. Jervis et al. [30] developed a lexicon consisting of 36 terms that described the diversity in crust and crumb appearance, flavor, hand texture, and oral texture characteristics of whole-wheat sandwich bread. Cho et al. [31] defined and referenced 27 terms for characterizing the flavor and texture properties of *seolgitteok* (Korean rice cake) formulated with varying levels of brown rice flour and sugar.

2.5. Dairy Products

Newman et al. [32] developed a sensory lexicon for dairy protein hydrolysates produced from whey protein and casein substrates with varying degrees of hydrolysis. The lexicon consisted of 19 flavor attributes, among which bitter taste, metallic, and astringent were important characteristics for differentiating casein hydrolysates from whey protein hydrolysates. Brown and Chambers [33] established a lexicon that described flavor and texture differences among yogurt samples of varying milk sources (organic or conventional), percent milk fat and processing (set, stirred, or strained/Greek styles). A lexicon for artisan goat milk cheese was developed using 47 samples manufactured in different parts of the U.S., and it consisted of 39 flavor terms [34]. Twenty-eight of the terms were commonly present in goat milk cheeses and were able to describe most flavor characteristics of the samples. In addition, common attributes that described goat milk cheese of certain types (chèvre-style, feta-style, cheddar-style, and mold ripened type) were listed.

2.6. Soy Products

Recent examples of lexicons for soy products included those for soy sauce, soy milk, *sufu* (fermented soybean curd, a side-dish or condiment of traditional Chinese cuisine) and *doenjang* (Korean fermented soybean paste). Cherdchu et al. [35] examined 20 representative soy sauce samples produced from several Asian regions and the U.S., and developed a lexicon consisting of 58 terms for describing the diversity in flavor characteristics of the soy sauce category. Since the soy sauce lexicon was developed based on an agreement between U.S. and Thai descriptive panels, the attribute terms were given in both English and Thai. The soy sauce lexicon of Cherdchu et al. [35] was expanded further in two later studies. Pujchakarn et al. [1] developed a lexicon consisting of 50 attributes for describing the appearance, aroma, flavor, and aftertaste of seasoning soy sauce, a specific subcategory of soy sauce. Seasoning soy sauce is mainly produced by a chemical hydrolysis process with an addition of other ingredients such as sugar, vinegar, and flavor enhancers. Many terms in the seasoning soy sauce lexicon were found in the general soy sauce lexicon of Cherdchu et al. [35], while seven new flavor attributes ("dark fruity", "celery", "Chinese radish", "fermented soybean", "malt/cereal", "musty", "prickly") were detected. Imamura [36] evaluated 149 soy sauce samples of varying manufacturing countries and methods and developed an exhaustive list of 88 terms for describing the aroma, flavor, and texture characteristics of the samples. The researchers also organized all attributes into a flavor wheel for a better understanding of the soy sauce characteristics. In addition, they reported that 19 out of the 88 attributes were common characteristics present in all soy sauce samples, and that evaluation of the 19 attributes was adequate for distinguishing naturally brewed soy sauce from chemically hydrolyzed soy sauce. Lawrence et al. [37] developed a lexicon for unflavored soymilk based on the evaluation of 26 commercial soymilk samples available in the U.S. market. The lexicon contained 24 terms, most of which described the flavor characteristics of soymilk. For *sufu*, lexicons were developed for the plain type [38] and red type [39], based on the evaluation of commercial *sufu* samples. The plain *sufu* lexicon consisted of 22 terms, including aroma, flavor, and texture attributes, among which "salty", "moldy", "alcohol-like", "sesame-like", and "cohesiveness" were deemed important for sample differentiation. The red *sufu* lexicon contained 15 descriptors for appearance, aroma, flavor, texture, and aftertaste characteristics. Kim et al. [40] defined and referenced 31 terms that described the appearance, aroma, flavor, texture, and aftertaste of *doenjang*, as affected by microbial communities present in the sample.

2.7. Meat Products

Baker et al. [41] developed a lexicon for sensory evaluation of caviar. The lexicon consisted of 18 descriptors with which to describe differences in appearance, aroma, flavor, and texture among caviar samples harvested from sturgeon-fed varying diets. Kim et al. [42] identified and referenced 18 terms for describing the appearance and flavor of chicken stock of varying forms (i.e., cube, liquid, and powder). Samant et al. [43] developed a lexicon consisting of 28 terms for describing the effects of smoking and marination on the aroma, flavor, and texture characteristics of chicken breast fillets. Lexicons for some traditional meat products, such as *salama da sugo* (a typical fermented sausage produced in Italy), Lucanian dry-cured sausage (Italian pork-based sausage), *morcela de Arroz* (Portuguese cooked blood sausage), bulgogi (Korean barbecued beef), and *larou* (Chinese traditional bacon) were also published. Coloretti et al. [44] developed a lexicon for the sensory evaluation of *salama da sugo*. The lexicon consisted of 23 descriptors, including "wine aroma", "spicy aroma", "astringent", "pricking", "fat/lean connection", and "fibrosity", among others. The lexicon for Lucanian dry-cured sausage consisted of 21 terms that described sensory characteristics of the product, as affected by the presence or absence of nitrates and nitrites as curing agents [45]. For Portuguese cooked blood sausage, 14 descriptors were defined and referenced [46]. The bulgogi lexicon listed 17 terms for describing the aroma, flavor, and texture of the product, as affected by varying levels of sugar, soy sauce, garlic, or sesame oil [47]. The terms were given in English and Korean. Wang et al. [48] identified and referenced 17 descriptors for describing a wide range of aroma and flavor characteristics of commercial *larou* samples produced from different geographic regions of China. The descriptors were given in English and Chinese.

2.8. Foods for Animals

To date, only two sensory lexicons for animal foods have been published in recent years, one for dog food and one for cat food. A lexicon for dry dog food was published by Di Donfrancesco et al. [49] who examined 21 representative dry dog food samples that were selected from approximately 200 samples commonly sold in the U.S., then defined and referenced 72 terms for describing the appearance, aroma, flavor, and texture of the samples. More recently, Koppel and Koppel [50] developed an aroma lexicon for retorted cat food. A total of 30 attributes were defined and referenced. The main aroma attributes in retorted cat food included "meaty", "brothy", "cooked", "vitamin", and "barnyard", and depending on the ingredients used, also included "poultry", "beefy liver", "seafood", and "heated oil" aromas.

2.9. Miscellaneous Items

Rosales et al. [51] defined and referenced 26 terms to describe the appearance, flavor, and texture of dark compound chocolate, as affected by the addition of crystal promoter additives. Jaffe et al. [52] explored smoky characteristics and developed a lexicon to describe the smoky flavor of a wide variety of smoked products, such as sausage, bacon, chicken, turkey, and ham lunch meat, marinade, barbeque sauce, cheese, fish, and liquid smoke. Fourteen attributes were defined and referenced, including aroma terms such as "smoky (overall)", "ashy", "woody", "burnt", "acrid", "pungent", "petroleum-like", "creosote/tar", "cedar", and "bitter", among others. Chambers et al. [53] defined and referenced 21 descriptors that described the texture properties of thickened foods during ingestion, swallowing, and after swallowing. The descriptors could be used to evaluate and compare the texture characteristics of thickened food products prescribed for patients with dysphagia.

For ease of reference, Table 1 lists the recent lexicons mentioned above, with details including: (1) the number of samples being examined for lexicon development, (2) whether or not the definition, reference, and reference intensity are given in the lexicons, and (3) the number of descriptors listed in the lexicons. It is recommended that the number of samples being examined during lexicon development should cover the entire product space [2]. The appropriate size of the sample set depends on the diversity of the product category, as well as the objective of the study. For studies that aim to develop a lexicon for describing sensory differences among samples as affected by certain factors varying at certain levels, a not very large sample set could perhaps represent the entire product space. This was the case for some of the studies shown in Table 1, in which the lexicons were developed and subsequently used to determine samples' sensory characteristics as affected by cultivars [9], cultivars/varieties and processing/preparation methods [12,24,25,43], and ingredients/additives [29,31,45,47,51] that were varied at limited levels. The number of samples being examined in those studies varied from 4 to 15. On the other hand, the rest of the studies in Table 1 aimed to develop general lexicons for specific product categories; therefore, a larger sample set was used in most studies in order to represent each entire product space. For instance, 105 coffee samples from countries around the world were examined to develop a general lexicon for brewed coffee [26]. Apple samples of 78 cultivars were used to develop a general lexicon for apple [6]. Forty retorted cat food samples varying in processing, ingredients, and packaging that represented the product market space were examined to develop a general lexicon for retorted cat food [50]. Twenty representative soy sauce samples produced in different countries were used to develop a general lexicon for soy sauce [35]. However, there were some studies [10,13–17,23,28,38–40,42,44,46] that used quite a small number of samples (n = 5–14) to develop a general lexicon for a specific product category. Those few samples were less likely to be able to fairly represent the entire product space. Consequently, the lexicons developed in those studies should be considered as "initial" lexicons, meaning that more attributes may become apparent in future studies with more varieties of samples.

Table 1. Examples of recent food lexicons.

	Number of Samples Used in Lexicon Development	Definitions	Character References	Reference Intensities	Appearance	Aroma	Flavor	Texture/Mouthfeel	Aftertaste	Source, Published Date
Fruits and Vegetables										
Apple	40	Yes	Yes	Yes	2	1	5	7	-	[5], 2013
Apple	78	Yes	Yes	No	-	-	12	6	-	[6], 2018
Pomelo	20	Yes	Yes	Yes	1	7	10	9	3	[7], 2015
Peach	51	Yes	Yes	No	1	11	14	7	-	[8], 2017
Blackberry	9	Yes	Yes*	Yes	-	17	-	-	-	[9], 2010
Strawberry	12	Yes	Yes	No	2	8	10	5	-	[10], 2018
Pomegranate	20	Yes	Yes	Yes	1	-	26	8	-	[11], 2014
Mango	12	Yes	Yes	Yes	-	-	30	10	-	[12], 2014
Sweet tamarind	12	Yes	Yes	Yes	-	-	17	9	-	[13], 2010
Cabbage kimchi	14	Yes	Yes	Yes	-	-	15	2	-	[14], 2012
Dried fig	12	No	Yes	No	13	20	23	7	5	[15], 2013
Grains and nuts										
Black walnut	7	Yes	Yes	Yes	-	-	22	-	-	[6], 2013
Black walnut	10	Yes	Yes	Yes	3	8	23	6	-	[17], 2016
Pecan	32	Yes	Yes	Yes	-	-	20	-	-	[18], 2016
Cashew nut	15	Yes	Yes	No	-	-	25	4	-	[19], 2017
Quinoa	21	Yes	Yes	Yes	3	9	7	8	-	[20], 2017
Spaghetti	50	Yes	Yes	Yes	5	9	2	19	-	[21], 2018
Beverages										
Blueberry juice	20	Yes	Yes	No	-	27 (aroma/flavor)		5	-	[22], 2013
Orange juice	7	Yes	Yes	No	-	-	15	8	-	[23], 2013
Coffee	9	Yes	Yes	Yes	-	15	-	-	-	[24], 2011
Brewed coffee	12	Yes	Yes	Yes	-	10	15	-	3	[25], 2015
Brewed coffee	105	Yes	Yes	No	-	110 (aroma and flavor)		2	-	[26], 2016
Hibiscus tea	22	Yes	Yes	No	4	15 (aroma and flavor)		2	-	[27], 2017
Pink port wine	5	Yes	Yes	Yes	3	7	6	4	1	[28], 2014
Bakery products										
Prebiotic gluten free bread	6	Yes	Yes	Yes	3	4	3	5	-	[29], 2014
Whole wheat sandwich bread	25	Yes	Yes	Yes*	5	-	19	12	-	[30], 2016
Seolgitteok (Korean rice cake)	15	Yes	Yes	Yes	-	-	12	15	-	[31], 2014
Dairy products										
Dairy protein hydrolysate	25	No	Yes	Yes	-	-	19	-	-	[32], 2014
Yogurt	29	Yes	Yes	Yes	-	-	25	10	-	[33], 2015
Goat milk cheese	47	Yes	Yes	Yes	-	-	39	-	-	[34], 2016
Soy products										
Soy sauce	20	Yes	Yes	Yes	-	-	58	-	-	[35], 2013
Seasoning soy sauce	25	Yes	Yes	Yes	1	22	24	-	3	[1], 2016
Soy sauce	149	Yes	Yes	No	-	47	37	4	-	[36], 2016
Soymilk	26	Yes	Yes	Yes*	1	1	21	1	-	[37], 2016
Plain *sufu* (Chinese fermented soybean curd)	12	Yes	Yes	Yes	-	8	7	7	-	[38], 2016
Red *sufu* (Chinese fermented soybean curd)	12	Yes	Yes	Yes	1	4	5	4	1	[39], 2018
Doenjang (Korean fermented soybean paste)	14	Yes	Yes	No	3	8	10	5	5	[40], 2016

Table 1. Cont.

	Number of Samples Used in Lexicon Development	Definitions	Character References	Reference Intensities	Appearance	Aroma	Flavor	Texture/Mouthfeel	Aftertaste	Source, Published Date
Fruits and Vegetables										
Meat products										
Caviar	55	Yes	Yes	Yes	6	1	7	4	-	[41], 2014
Chicken stock	10	Yes	Yes	Yes	5	-	13	-	-	[42], 2017
Chicken breast filet	4	Yes	Yes	Yes	-	10	12	6	-	[43], 2016
Salama da sugo (Italian fermented sausage)	12	Yes	Yes	Yes	2	3	10	8	-	[44], 2015
Lucanian dry sausage (Italian pork-based sausage)	10	Yes	Yes	Yes	7	4	6	4	-	[45], 2016
Morcela de Arroz (Portuguese cooked blood sausage)	12	Yes	Yes	No	4	3	3	4	-	[46], 2015
Bulgogi (Korean traditional barbecued beef)	4	Yes	Yes	Yes	-	5	10	2	-	[47], 2011
Larou (Chinese traditional bacon)	24	Yes	Yes	Yes	-	17 (aroma/flavor)		-	-	[48], 2018
Animal foods										
Dry dog food	21	Yes	Yes	Yes	16	44 (aroma/flavor)		12	-	[49], 2012
Retorted cat food	40	Yes	Yes	Yes	-	30	-	-	-	[50], 2018
Miscellaneous										
Compound chocolate	8	Yes	Yes	Yes	2	-	13	11	-	[51], 2018
Smoke attribute	54	Yes	Yes	Yes	-	-	14	-	-	[52], 2017
Thickened liquids	40	Yes	Yes	Yes	-	-	-	21	-	[53], 2017

* Reference intensities are given for some attributes.

When a large number of samples are used to develop a lexicon, samples are often placed in different sets to be used at different phases of lexicon development. For example, Chambers et al. [26] placed 105 coffee samples in four sets (n=13, 45, 27, 20 for sets 1–4, respectively). Set 1, with a narrow range of coffee samples, was used for initial lexicon development. Subsequently, Set 2, with an array of commercial coffee samples, and Set 3, with samples having unique sensory characteristics, were used to expand the lexicon and incorporate additional attributes. Lastly, Set 4, with a narrow range of coffee samples, was used to validate that the developed lexicon could explain diverse sensory characteristics present in all samples and, at the same time, could detect small differences among the samples. Another approach of sample arrangement was adopted in a study of Cherdchu et al. [35] in which a total of 116 soy sauce samples were procured from different manufacturing regions and initially screened by sensory analysts; thereafter, only 20 samples that represented the diversity in flavor characteristics of soy sauce were used for developing a soy sauce lexicon.

It should be noted that definitions and character references were provided for each descriptor in almost all of the lexicons listed in Table 1. According to Drake and Civille [4], the definition and character reference are crucial components of a sensory lexicon, since they help to maximize language clarity and enable panelists and other audiences to clearly understand the concept of each term. However, there was one study [32] that only provided the character references without a term definition. Most of the lexicons in Table 1 also provided the intensity of each reference to anchor an attribute scale that could help reduce panel variation in the intensity rating.

3. Applications of Lexicons

Applications of sensory lexicons as an effective communication tool and a guidance tool for research and development of food products are as follows.

3.1. Communication Tool

Lexicons facilitate accurate and precise communication regarding products' sensory characteristics across diverse audiences, such as sensory panels and scientists, product developers, marketing professionals, and suppliers, both within and between companies and even between countries [2,21]. Different people, especially those from different countries and cultures, may have a different understanding and interpretation of the same descriptor. However, this deviation could possibly be solved with the use of a well-developed lexicon in which definition and character reference(s) for each sensory descriptor are provided. For example, Cherdchu et al. [35] found that U.S. and Thai panels had difficulties understanding some of the other panel's terms for describing the flavor characteristics of soy sauce. Thai panelists had difficulty differentiating the terms "brown", "caramel", and "dark brown" that were used by U.S. panelists. Both panels had difficulties differentiating the terms "earthy-damp", "dusty", and "moldy-damp". However, the problems were solved once the meaning of each term was clearly defined with an agreement of both panels and an appropriate reference standard was used to represent each term. This study highlighted the importance of term definitions and reference standards for cross-cultural sensory research. A study by Vázquez-Araújo et al. [54] demonstrated that a lexicon for *turrón* (European confectionary product) that was developed by the U.S. panel could be adopted effectively by the Spanish panel for evaluating the product. They found that flavor attributes of *turrón* samples were rated in the same way by both panels, suggesting a similar understanding toward flavor attributes of the two panels. However, there were inconsistencies between the two panels on some texture attributes. The authors suggested that definitions and references of these texture attributes should be adapted.

Some terms describe complex attributes which may exert a different character depending on products in which they are present. For example, both spinach and cucumber could be described as "green"; however, the "green" character of spinach is different from that of cucumber. Therefore, it would be advantageous if ones could describe any products using more specific terms. Hongsoongnern and Chambers [55] explored the "green" character of various food categories. They found that the "green" character could be described using five terms: "green-unripe", "green-peapod", "green-grassy/leafy", "green-viney", and "green-fruity", and food examples of these green related terms were green banana, raw peanut, spinach, cucumber, and green apple, respectively. In addition to the "green" attribute, research had been carried out to develop lexicons for describing other complex sensory attributes, including "smoky" [52], "nutty" [56], and "beany" [57,58]. This style of lexicon could be useful for communication among various audiences.

3.2. New Product Development

For the food industry, new product development is one of their most important activities. The process of developing new products can be divided into several stages, among which concept development, prototyping, and commercialization are critical steps to the success of a new product [59]. Descriptive tests and sensory lexicons developed by trained panelists could be used as effective guidance tools in a new product development process as follows.

3.2.1. Concept Development

For concept development, information is gathered from the literature, patents, market trends, and competitive products in the market. Descriptive sensory analysis could provide better understanding of sensory properties of existing or competitive products where the potential new product will be placed [60]. A number of lexicons were developed and used to determine sensory profiles of commercial products, such as cabbage kimchi [14], hibiscus tea [27], goat milk cheese [34], French cheese [61], pomegranate juice [62], green tea [63], plain *sufu* (Chinese fermented soybean curd) [38], dry dog food [49], and retorted cat food [50], among others.

In addition, combining descriptive data with consumer acceptability data could assist in identifying which products are liked for which reason, and possibly providing insight on potential gaps in the market [60]. For instance, Lawrence et al. [37] studied sensory profiles of 26 commercial plain soymilks available in the U.S. market using a lexicon consisting of 24 terms, and subsequently selected 12 representative soymilk samples based on their sensory profiles for testing with consumers from different age/cultural categories. Based on liking data, three consumer clusters were identified. Consumers in Cluster 1 were mainly Asian females aged 18–30 years, while those in Cluster 2 were mainly Caucasian/African American females aged 40–64 years. Cluster 3 consisted of Caucasian/African American females aged 18–30 and 40–64 years. With a preference mapping technique, the drivers of liking for each consumer cluster were identified. Sweet taste, sweet aromatic, vanilla/vanillin flavor, and higher viscosity were preferred by all consumer clusters, and differences among the clusters were the drivers of disliking. There was no driver of disliking identified for Cluster 1, while Clusters 2 and 3 disliked soymilk with "beany", "green/grassy", and "meaty/brothy" flavors and astringency. The consumers in Cluster 3 were willing to overlook disliked attributes with the addition of sweet taste, while those in Cluster 2 were not.

In another study, Lykomitros et al. [64] reported that roasted peanut, dark roast and sweet aromas, and sweet taste were the drivers of liking, while raw bean aroma and bitter taste were detractors for the liking of roasted peanut for European consumers. Kim et al. [23] studied sensory profiles of commercial orange juice products in the Korean market using a lexicon with 23 descriptors, and identified drivers of liking of Korean consumers. They found three consumer segments based on liking. Consumers in Segment 1 (46%) preferred orange juice products with strong orange flavor, which could be either artificial or natural. Consumers in Segment 2 (29%) were inclined to "functional flavors", such as "sour", "bitter", and "medicinal", while those in Segment 3 (25%) liked orange juice products that were high in sweet taste but low in sour and bitter tastes.

Studies on drivers of liking and disliking of other food products, such as gluten-free bread [29], whole wheat sandwich bread [30], *doenjang* (Korean fermented soybean paste) [40], caviar [41], and sweet potato [65] were also available in the literature. An understanding of how sensory properties affect consumer preferences is very important to product developers to create products with attributes that are well-liked by target consumers and to tailor attributes for segments of the population that have not yet been accommodated [37]. A profile of desirable sensory characteristics not only aids the development of products, but also serves as a reference guide once prototypes are being developed [60].

3.2.2. Prototyping

Carrying out descriptive analyses using sensory lexicons throughout the development stage is advantageous for product developers, as it helps them to understand the effects of ingredients, processing methods, packaging conditions, etc. on the sensory qualities of products in detail. In addition, the developed products should be compared with desirable sensory characteristics identified in the product concept [60]. For instance, Cho et al. [31] used a lexicon consisting of 27 descriptors to determine the effects of the amount of brown rice flour and sugar on flavor and texture properties of *seolgitteok* (Korean rice cake). The *bulgogi* lexicon, with five aroma, ten flavor, and two texture terms was used to assess the effects of sugar, soy sauce, garlic, and sesame oil on the sensory characteristics of the product [47]. A lexicon developed for dairy protein hydrolysates (casein and whey protein hydrolysates) could be useful for developing products using these ingredients as protein sources, such as baby foods, nutrient-rich beverages, sport drinks, etc. [32]. In particular, attention should be paid to the attributes in the lexicon—namely, "sulfurous", "wet dog/animal", "metallic", "bitter", and "astringent", since they are undesirable characteristics caused by dairy protein hydrolysates. Ledeker et al. [66] conducted a descriptive test using a lexicon with 46 descriptors to understand the effects of heating on the flavor and texture qualities of mango purée and sorbet, prepared from four mango cultivars available in the U.S. Later, a similar study was conducted on mango purée prepared from Thai mango cultivars [12]. Jáuregui et al. [67] used a lexicon containing 13 descriptors to understand the effects of different vegetable oils on the sensory quality and stability of fried salted soybean.

Once the product prototype is developed, a sensory lexicon could be used for comparing the sensory property of the prototype against that of commercial products, especially the category leaders, to ensure that the prototype will be successful in the market. For example, Brown and Chambers [33] compared the sensory characteristics of two prototype yogurts against 26 commercial plain yogurts. The prototypes differed from commercial plain yogurts in that they had undergone an additional pre-acidification step, in which milk was pre-acidified with lemon juice or citric acid to pH 6.2 prior to the normal fermentation process in order to shorten fermentation time and thus reduce production energy. Highly trained descriptive panelists evaluated all yogurt samples using a lexicon consisting of 35 attributes. Results showed that flavor and texture characteristics of the prototypes were comparable to those of leading brands, thereby suggesting potential market viability of the prototypes.

3.2.3. Commercialization

Product commercialization is a full scale-up and integration of production and marketing [59]. Scaling-up from the test kitchen to the pilot plant to the factory trail always produces changes in a product. Performing descriptive tests using a lexicon could assist in understanding subtle sensory changes during a scale-up, and could help find solutions for bringing the product back to the originally intended sensory characteristics [60].

3.3. Quality Control

The role of sensory evaluation in quality control usually involves the maintenance of consistency of food quality [60]. Descriptive sensory analysis and lexicons could be used to determine the target specification of food products in terms of sensory quality, and to test if the product complies with the target specification [68]. In setting quality specifications, the relationship between consumer acceptance and descriptive panel responses should be established, and a method to determine such a relationship was described in detail by Moskowitz et al. [69]. A study by Chambers et al. [14] demonstrated a possible use of lexicons for quality control purposes. In their study, a lexicon consisting of 17 terms were used to evaluate flavor and texture characteristics of Korean commercial cabbage kimchi (including wedge vs.

sliced types, and fresh vs. fermented types). Results showed that there were products of the same type from the same manufacturer with totally different sensory characteristics. Since they were commercial products, it was not known with certainty whether that deviation was due to a poor quality-control system, or whether it was because the manufacturer had intentionally different products to appeal to different segments of the consumer market. In the case where an improved quality-control system was needed, the lexicon developed in the study could be used to pinpoint different characteristics and then adjust the production process to improve product consistency.

3.4. Product Improvement

Descriptive sensory analyses and lexicons could be used to monitor whether changes in ingredients, processing methods, and packaging could improve the quality of a company's existing products. For example, a study by Rosales et al. [51] used a descriptive test and a lexicon containing 25 well-defined and referenced terms to determine if heat-resistant properties and the sensory quality of dark-compound chocolate could be improved with the addition of a crystal-promoter additive. Three crystal-promoter additives were tested at varying concentrations. Results showed that the additive CP1 (at 0.25% level) which was composed of mono- and diglycerides and polyglycerol esters from high oleic sunflower oil was better than the other two additives at all concentration levels, resulting in compound chocolate with high heat-resistant properties and with higher cocoa, dark brown, bitter aromatic, and sweet intensities, as well as a faster melting rate and less wax-coating mouthfeel.

3.5. Measuring Changes during Product Shelf-Life

ASTM International [70] defined sensory shelf-life as "the time period during which the products' sensory characteristics and performance are as intended by the manufacturer". Descriptive tests and lexicons developed by trained panelists are effective tools in shelf-life studies. Such tools could be used to track changes in the sensory characteristics of a product over time, or to determine how long a product can be stored before there is a noticeable change in sensory quality [68].

For instance, Rosales and Suwonsichon [7] monitored changes in the sensory quality of five pomelo cultivars during seven-day fresh-cut storage (4 °C, 85% relative humidity) using a lexicon consisting of 30 attributes established by a highly trained descriptive panel. They found that pomelo samples of Tubtim Siam, Thong Dee, and Kao Numpeung cultivars showed only minimal changes throughout the storage period. While the Kao Tangkwa cultivar showed noticeable decreases in hardness, firmness, chewiness, and sourness, it showed an increase in pomelo identity and overall sweet notes after storage for three days. For Kao Yai cultivar, changes in sensory quality became more evident after storage for five days, resulting in increased hardness and firmness but decreased sweetness. Such information could be useful for retailers in regard to knowing the approximate storage duration of the pre-cut pomelo fruit of different cultivars without noticeable quality changes.

Riveros et al. [71] compared sensory profiles of peanut paste prepared from high-oleic and normal peanuts. A lexicon consisting of 16 attributes was developed for tracking the sensory changes of the samples during storage at 4, 23, and 40 °C for up to 175 days. Results showed significant changes in three attributes, with peanut paste prepared from high-oleic peanut having increases in oxidized and cardboard flavors and a decrease in roasted peanut flavor at slower rates than that prepared from normal peanuts. In another study of the same researchers, changes in sensory attributes of roasted peanuts coated with different edible coatings (carboxymethyl cellulose, methyl cellulose, and whey protein) were monitored during storage at 40 °C for up to 56 days [72]. Trained panelists rated intensities of 13 attributes selected from the peanut lexicon of Johnsen et al. [73], three of which changed significantly during storage. Roasted peanut flavor was lower, while oxidized and cardboard flavors were higher for uncoated peanuts than for coated ones. Among the edible coatings, carboxymethyl cellulose provided the best protection effect.

Lee and Chambers [74] determined changes in flavor characteristics of two Korean green teas during storage at 20 °C for 3, 6, 12, 18, and 24 months. Twenty flavor attributes selected from the green tea lexicon developed earlier by the same authors [75] were evaluated. At each storage period, panelists rated attribute intensities of the samples by comparing them to the reference standard scores provided in the lexicon. The authors pointed out that comparison to reference standards was necessary to ensure that the panelists did not drift in their assessments over the two-year testing period. This study highlighted the importance of quantitative references in the lexicon for long-term storage testing. Results revealed that both green tea samples changed only slightly over a period of two years.

3.6. Breeding New Plant Cultivars

Plant breeding has been practiced for thousands of years to change the traits of plants in order to produce the desired characteristics [76]. Understanding sensory characteristics and consumer preference is an important key to develop and select novel plant cultivars that will be successful in the marketplace [6]. Lexicons have been widely used for descriptive tests to identify, quantify, and compare sensory characteristics of fruits and vegetables among cultivars and to relate descriptive data to instrumental data and consumer liking to aid in breeding new successful cultivars. Lexicons established for fruits, vegetables, and nuts which could be used for this purpose are prominent in the literature, some of which are shown in Table 1.

The application of sensory lexicons in plant breeding can be demonstrated through the studies of Bowen et al. [6], Du et al. [9], and Vara-Ubol et al. [77]. Bowen et al. [6] determined sensory profiles of 78 apple cultivars using a lexicon established by a trained descriptive panel. Nineteen representative apple cultivars were then subjected to an acceptance test using 219 consumers. Afterward, relationships between sensory profile data and liking data were determined through the external preference mapping technique. Results showed that apple cultivars could be classified into four groups based on the following flavor and texture properties: "aromatic-sweet", "acidic", "mealy", and "balanced". Most consumers (89%) preferred apple cultivars that were sweet and which had a fresh red-apple aroma, while the rest (11%) liked apple cultivars that had an acidic taste and green-apple aroma. For all consumers, a high level of perceivable crispy and juicy texture was a strong driver of liking, whereas a mealy texture was a strong driver of disliking. Such information, as well as an external preference map created in this study, could be useful for selecting apple cultivars with desired sensory characteristics for crossing in a breeding program.

In Du et al.'s study [9], flavor characteristics of eight thornless blackberry cultivars were compared against those of Marion, a thorny cultivar. Although Marion is the most commonly marketed cultivar, its thorns can be dangerous to consumers if they end up in the fruit. Therefore, the aim of the study was to identify thornless cultivars that had similar or superior flavor quality to the Marion cultivar. The thornless blackberry cultivars being tested included six commercial cultivars (Thornless Evergreen, Black Diamond, Black Pearl, Nightfall, Waldo, and Chester) and two newly bred cultivars (ORUS 1843-3 and NZ 9351-4). Trained descriptive panelists evaluated all of the blackberry cultivars using a lexicon consisting of 17 flavor attributes. Results showed that four (Black Diamond, Black Pearl, ORUS 1843-3, and NZ 9351-4) out of eight thornless cultivars were similar to the Marion, having high intensities of "fresh-fruity", "raspberry", "floral", and "strawberry" flavors, while the rest of the thornless cultivars had high intensities of "vegetal", "woody", "moldy", and "cooked fruit" flavors. Results of this study could be provided as a guide to blackberry breeding programs to develop or select thornless cultivars with superior fruit quality.

Vara-Ubol et al. [77] determined sensory characteristics of eight rose-apple varieties cultivated in Thailand, five of which (Pet-Num-Ping, Toon-Klao, Tub-Tim-Jun, Pet-Sai-Rung, and Pet-Sam-Pran) were popular in the marketplace, while three of which (Ma-Meow, Num-Dok-Mai, and Sa-Rak) were not. Results showed distinct differences in sensory characteristics between the popular and unpopular cultivars. The former had a crispy texture but bland flavor, while the latter had a spongy, mealy, and firm texture with strong fruity, floral/perfumy, rose, and woody flavor notes. The authors pointed out the possibility that the cultivars Pet-Num-Ping, Toon-Klao, Tub-Tim-Jun, Pet-Sai-Rung, and Pet-Sam-Pran could become even more accepted by consumers if plant breeders could combine the crispy texture of the cultivars with the rose and floral flavor notes.

4. Conclusions

A number of recent sensory lexicons developed by trained panelists for documenting and describing sensory characteristics of various food products have been provided in the current review. This review has also shown how the lexicons can be used as an effective communication tool and research guidance tool for new product development processes, quality control, product improvement, measuring changes of products during storage, and breeding new plant cultivars.

Funding: This research received no external funding.

Acknowledgments: The author wish to thank Edgar IV Chambers, Center for Sensory Analysis and Consumer Behavior, Kansas State University, for his constructive guidance on the manuscript.

Conflicts of Interest: The authors declare no conflict of interest.

References

1. Pujchakarn, T.; Suwonsichon, S.; Suwonsichon, T. Development of a sensory lexicon for a specific subcategory of soy sauce: Seasoning soy sauce. *J. Sens. Stud.* **2016**, *31*, 443–452. [CrossRef]
2. Lawless, L.J.R.; Civille, G.V. Developing lexicons: A review. *J. Sens. Stud.* **2013**, *28*, 270–281. [CrossRef]
3. Mozkowitz, H.R.; Muñoz, A.M.; Gacula, M.C., Jr. (Eds.) Language development in descriptive analysis and the formation of sensory concepts. In *Viewpoints and Controversies in Sensory Science and Consumer Product Testing*; Food & Nutrition Press, Inc.: Trumbull, CT, USA, 2003; pp. 313–336. ISBN 0-9176-7857-5.
4. Drake, M.A.; Civille, G.V. Flavor lexicons. *Compr. Rev. Food Sci. Food Saf.* **2002**, *2*, 33–40. [CrossRef]
5. Corollaroa, M.L.; Endrizzia, I.; Bertolini, A.; Apreaa, E.; Luisa Dematté, M.; Costaa, F.; Biasioli, F.; Gasperia, F. Sensory profiling of apple: Methodological aspects, cultivar characterisation and postharvest changes. *Postharvest Biol. Technol.* **2013**, *77*, 111–120. [CrossRef]
6. Bowen, A.J.; Blake, A.; Tureček, J.; Amyotte, B. External preference mapping: A guide for a consumer-driven approach to apple breeding. *J. Sens. Stud.* **2018**, e12472. [CrossRef]
7. Rosales, C.K.; Suwonsichon, S. Sensory lexicon of pomelo fruit over various cultivars and fresh-cut storage. *J. Sens. Stud.* **2015**, *30*, 21–32. [CrossRef]
8. Belisle, C.; Adhikari, K.; Chavez, D.; Phan, U.T.X. Development of a lexicon for flavor and texture of fresh peach cultivars. *J. Sens. Stud.* **2017**, *32*, e12276. [CrossRef]
9. Du, X.F.; Kurnianta, A.; McDaniel, M.; Finn, C.E.; Qian, M.C. Flavour profiling of 'Marion' and thornless blackberries by instrumental and sensory analysis. *Food Chem.* **2010**, *121*, 1080–1088. [CrossRef]
10. Oliver, P.; Cicerale, S.; Pang, E.; Keast, R. Developing a strawberry lexicon to describe cultivars at two maturation stages. *J. Sens. Stud.* **2018**, *33*, e12312. [CrossRef]
11. Vázquez-Araújo, L.; Nuncio-Jáuregui, P.N.; Cherdchu, P.; Hernández, F.; Chambers, E., IV; Carbonell-Barrachina, Á.A. Physicochemical and descriptive sensory characterization of Spanish pomegranates: Aptitudes for processing and fresh consumption. *Int. J. Food Sci. Technol.* **2014**, *49*, 1663–1672. [CrossRef]
12. Ledeker, C.N.; Suwonsichon, S.; Chambers, D.H.; Adhikari, K. Comparison of sensory attributes in fresh mangoes and heat-treated mango purées prepared from Thai cultivars. *LWT Food Sci. Technol.* **2014**, *56*, 138–144. [CrossRef]
13. Oupadissakoon, C.; Chambers, E., IV; Kongpensook, V.; Suwonsichon, S.; Yenket, R.; Retiveau, A. Sensory properties and consumer acceptance of sweet tamarind varieties grown in Thailand. *J. Sci. Food Agric.* **2010**, *90*, 1081–1088. [CrossRef] [PubMed]
14. Chambers, E., IV; Lee, J.; Chun, S.; Miller, A.E. Development of a lexicon for commercially available cabbage (baechu) kimchi. *J. Sens. Stud.* **2012**, *27*, 511–518. [CrossRef]
15. Haug, M.T.; King, E.S.; Heymann, H.; Crisosto, C.H. Sensory profiles for dried fig (*Ficuscarica* L.) cultivars commercially grown and processed in California. *J. Food Sci.* **2013**, *78*, S1273–S1281. [CrossRef] [PubMed]

16. Miller, A.E.; Chambers, D.H. Descriptive analysis of flavor characteristics for black walnut cultivars. *J. Food Sci.* **2013**, *78*, S887–S893. [CrossRef] [PubMed]
17. Lynch, C.; Koppel, K.; Reid, W. Sensory profiles and seasonal variation of black walnut cultivars. *J. Food Sci.* **2016**, *81*, S719–S727. [CrossRef]
18. Magnuson, S.M.; Kelly, B.; Koppel, K.; Reid, W. A comparison of flavor differences between pecan cultivars in raw and roasted forms. *J. Food Sci.* **2016**, *81*, S1243–S1253. [CrossRef]
19. Griffin, L.E.; Dean, L.L.; Drake, M.A. The development of a lexicon for cashew nuts. *J. Sens. Stud.* **2017**, *32*, e12244. [CrossRef]
20. Wu, G.; Ross, C.F.; Morris, C.F.; Murphy, K.M. Lexicon development, consumer acceptance, and drivers of liking of quinoa varieties. *J. Food Sci.* **2017**, *82*, 993–1005. [CrossRef]
21. Irie, K.; Maeda, T.; Kazami, Y.; Yoshida, M.; Hayakawa, F. Establishment of a sensory lexicon for dried long pasta. *J. Sens. Stud.* **2018**, e12438. [CrossRef]
22. Bett-Garber, K.L.; Lea, J.M. Development of flavor lexicon for freshly pressed and processed blueberry juice. *J. Sens. Stud.* **2013**, *28*, 161–170. [CrossRef]
23. Kim, M.K.; Lee, Y.J.; Kwak, H.S.; Kang, M. Identification of sensory attributes that drive consumer liking of commercial orange juice products in Korea. *J. Food Sci.* **2013**, *78*, S1451–S1458. [CrossRef] [PubMed]
24. Bhumiratana, N.; Adhikari, K.; Chambers, E., IV. Evolution of sensory aroma attributes from coffee beans to brewed coffee. *LWT Food Sci. Technol.* **2011**, *44*, 2185–2192. [CrossRef]
25. Sanchez, K.; Chambers, E., IV. How does product preparation affect sensory properties? An example with coffee. *J. Sens. Stud.* **2015**, *30*, 499–511. [CrossRef]
26. Chambers, E., IV; Sanchez, K.; Phan, U.X.T.; Miller, R.; Civille, G.V.; Di Donfrancesco, B. Development of a "living" lexicon for descriptive sensory analysis of brewed coffee. *J. Sens. Stud.* **2016**, *31*, 465–480. [CrossRef]
27. Monteiro, M.J.P.; Costa, A.I.A.; Franco, M.I.; Bechoff, A.; Cisse, M.; Geneviève, F.; Tomlins, K.; Pintado, M.M.E. Cross-cultural development of hibiscus tea sensory lexicons for trained and untrained panelists. *J. Sens. Stud.* **2017**, *32*, e12297. [CrossRef]
28. Monteiro, B.; Vilela, A.; Correia, E. Sensory profile of pink port wines: Development of a flavor lexicon. *Flavour Fragr. J.* **2014**, *29*, 50–58. [CrossRef]
29. Morais, E.C.; Cruz, A.G.; Faria, J.A.F.; Bolini, H.M.A. Prebiotic gluten-free bread: Sensory profiling and drivers of liking. *LWT Food Sci. Technol.* **2014**, *55*, 248–254. [CrossRef]
30. Jervis, S.M.; Guthrie, B.; Guo, G.; Worch, T.; Hasted, A.; Drake, M.A. Comparison of preference mapping methods on commodity foods with challenging groups of low-variance attributes: Sliced whole wheat sandwich bread example. *J. Sens. Stud.* **2016**, *31*, 34–49. [CrossRef]
31. Cho, S.; Yoon, S.H.; Min, J.; Tokari, S.T.; Lee, S.O.; Seo, H.S. Sensory characteristics of seolgitteok (Korean rice cake) in relation to the added levels of brown rice flour and sugar. *J. Sens. Stud.* **2014**, *29*, 371–383. [CrossRef]
32. Newman, J.; O'Riordan, D.; Jacquer, J.C.; O'Sullivan, M. Development of a sensory lexicon for dairy protein hydrolysates. *J. Sens. Stud.* **2014**, *29*, 413–424. [CrossRef]
33. Brown, M.D.; Chambers, D.H. Sensory characteristics and comparison of commercial plain yogurts and 2 new production sample options. *J. Food Sci.* **2015**, *80*, S2957–S2969. [CrossRef] [PubMed]
34. Talavera, M.; Chambers, D.H. Flavor lexicon and characteristics of artisan goat cheese from the United States. *J. Sens. Stud.* **2016**, *31*, 492–506. [CrossRef]
35. Cherdchu, P.; Chambers, E., IV; Suwonsichon, T. Sensory lexicon development using trained panelists in Thailand and the U.S.A.: Soy sauce. *J. Sens. Stud.* **2013**, *28*, 248–255. [CrossRef]
36. Imamura, M. Descriptive terminology for the sensory evaluation of soy sauce. *J. Sens. Stud.* **2016**, *31*, 393–407. [CrossRef]
37. Lawrence, S.E.; Lopetcharat, K.; Drake, M.A. Preference mapping of soymilk with different U.S. consumers. *J. Food Sci.* **2016**, *81*, S463–S476. [CrossRef]
38. Chen, Y.P.; Chung, H.Y. Development of a lexicon for commercial plain sufu (fermented soybean curd). *J. Sens. Stud.* **2016**, *31*, 22–33. [CrossRef]
39. He, W.; Chen, Y.P.; Chung, H.Y. Development of a lexicon for red sufu. *J. Sens. Stud.* **2018**, e12461. [CrossRef]
40. Kim, M.J.; Kwak, H.S.; Jung, H.Y.; Kim, S.S. Microbial communities related to sensory attributes in Korean fermented soy bean paste (doenjang). *Food Res. Int.* **2016**, *89*, 724–732. [CrossRef]

41. Baker, A.K.; Vixie, B.; Rasco, B.A.; Ovissipour, M.; Ross, C.F. Development of a lexicon for caviar and its usefulness for determining consumer preference. *J. Food Sci.* **2014**, *79*, S2533–S2541. [CrossRef]
42. Kim, H.; Lee, J.; Kim, B. Development of an initial lexicon for and impact of forms (cube, liquid, powder) on chicken stock and comparison to consumer acceptance. *J. Sens. Stud.* **2017**, e12251. [CrossRef]
43. Samant, S.S.; Crandall, P.G.; O'Bryan, C.A.; Lingbeck, J.M.; Martin, E.M.; Tokar, T.; Seo, H. Effects of smoking and marination on the sensory characteristics of cold-cut chicken breast filets: A pilot study. *Food Sci. Biotechnol.* **2016**, *25*, 1619–1625. [CrossRef] [PubMed]
44. Coloretti, F.; Grazia, L.; Gardini, F.; Lanciotti, R.; Montanari, C.; Tabanelli, G.; Chiavari, C. A procedure for the sensory evaluation of Salama da sugo, a typical fermented sausage produced in the Emilia Romagna Region, Italy. *J. Sci. Food Agric.* **2015**, *95*, 1047–1054. [CrossRef] [PubMed]
45. Braghieri, A.; Piazzolla, N.; Carlucci, A.; Bragaglio, A.; Napolitano, F. Sensory properties, consumer liking and choice determinants of Lucanian dry cured sausages. *Meat Sci.* **2016**, *111*, 122–129. [CrossRef]
46. Pereira, J.A.; Dionísio, L.; Matos, T.J.S.; Patarata, L. Sensory lexicon development for a Portuguese cooked blood sausage-*morcela de arroz de monchique*- to predict its usefulness for a geographical certification. *J. Sens. Stud.* **2015**, *30*, 56–67. [CrossRef]
47. Hong, J.H.; Yoon, E.K.; Chung, S.J.; Chung, L.; Cha, S.M.; O'Mahony, M.; Vickers, Z.; Kim, K.O. Sensory characteristics and cross-cultural consumer acceptability of bulgogi (Korean traditional barbecued beef). *J. Food Sci.* **2011**, *76*, S306–S313. [CrossRef] [PubMed]
48. Wang, H.; Zhang, X.; Suo, H.; Zhao, X.; Kan, J. Aroma and flavor characteristics of commercial Chinese traditional bacon from different geographical regions. *J. Sens. Stud.* **2018**, e12475. [CrossRef]
49. Di Donfrancesco, B.; Koppel, K.; Chambers, E., IV. An initial lexicon for sensory properties of dry dog food. *J. Sens. Stud.* **2012**, *27*, 498–510. [CrossRef]
50. Koppel, S.; Koppel, K. Development of an aroma attributes lexicon for retorted cat foods. *J. Sens. Stud.* **2018**, e12321. [CrossRef]
51. Rosales, C.K.; Suwonsichon, S.; Klinkesorn, U. Influence of crystal promoters on sensory characteristics of heat-resistant compound chocolate. *Int. J. Food Sci. Technol.* **2018**, *53*, 1459–1467. [CrossRef]
52. Jaffe, T.R.; Wang, H.; Chambers, E., IV. Determination of a lexicon for the sensory flavor attributes of smoked food products. *J. Sens. Stud.* **2017**, *32*, e12262. [CrossRef]
53. Chambers, E., IV; Jenkins, A.; Garcia, J.M. Sensory texture analysis of thickened liquids during ingestion. *J. Sens. Stud.* **2017**, *48*, 518–529. [CrossRef] [PubMed]
54. Vázquez-Araújo, L.; Chambers, D.H.; Carbonell-Barrachina, Á.A. Development of a sensory lexicon and application by an industry trade panel for turrón, a European protected product. *J. Sens. Stud.* **2012**, *27*, 26–36. [CrossRef]
55. Hongsoongnern, P.; Chambers, E., IV. A lexicon for green odor or flavor and characteristics of chemicals associated with green. *J. Sens. Stud.* **2008**, *23*, 205–221. [CrossRef]
56. Miller, A.E.; Chambers, E., IV; Jenkins, A.; Lee, J.; Chambers, D.H. Defining and characterizing the "nutty" attribute across food categories. *Food Qual. Pref.* **2013**, *27*, 1–7. [CrossRef]
57. Vara-Ubol, S.; Chambers, E., IV; Chambers, D.H. Sensory characteristics of chemical compounds potentially associated with beany aroma in foods. *J. Sens. Stud.* **2004**, *19*, 15–26. [CrossRef]
58. Bott, L.; Chambers, E., IV. Sensory characteristics of combinations of chemicals potentially associated with beany aroma in foods. *J. Sens. Stud.* **2006**, *21*, 308–321. [CrossRef]
59. Moskowitz, D.; Huang, Y.W. Indentifying critical steps in the new product development process. In *Accelerating New Food Product Design and Development*, 2nd ed.; Beckley, J.H., Herzog, L.J., Foley, M.M., Eds.; Wiley Blackwell: Hoboken, NJ, USA, 2017; pp. 249–258. ISBN 978-1-119-14930-9.
60. Beeren, C. Application of descriptive sensory analysis to food and drink products. In *Descriptive Analysis in Sensory Evaluation*; Kemp, S.E., Hort, J., Hollowood, T., Eds.; Wiley Blackwell: Hoboken, NJ, USA, 2018; pp. 611–646. ISBN 9780470671399.
61. Rétiveau, A.; Chambers, D.H.; Esteve, E. Developing a lexicon for the flavor description of French cheeses. *Food Qual. Pref.* **2005**, *16*, 517–527. [CrossRef]
62. Koppel, K.; Chambers, E., IV. Development and application of a lexicon to describe the flavor of pomegranate juice. *J. Sens. Stud.* **2010**, *25*, 819–837. [CrossRef]

63. Lee, J.; Chambers, D.H.; Chambers, E., IV. A comparison of the flavor of green teas from around the world. *J. Sci. Food Agric.* **2014**, *94*, 1315–1324. [CrossRef]
64. Lykomitros, D.; Fogliano, V.; Capuano, E. Drivers of preference and perception of freshness in roasted peanuts (*Arachis* spp.) for European consumers. *J. Food Sci.* **2018**, *83*, 1103–1115. [CrossRef] [PubMed]
65. Leksrisompong, P.P.; Whitson, M.E.; Truong, V.D.; Drake, M.A. Sensory attributes and consumer acceptance of sweet potato cultivars with varying flesh colors. *J. Sens. Stud.* **2012**, *27*, 59–69. [CrossRef]
66. Ledeker, C.N.; Chambers, D.H.; Chambers, E., IV; Adhikari, K. Changes in the sensory characteristics of mango cultivars during the production of mango pureé and sorbet. *J. Food Sci.* **2012**, *77*, S348–S355. [CrossRef] [PubMed]
67. Jáuregui, M.P.; Riveros, C.; Nepote, V.; Grosso, N.R.; Gayol, M.F. Chemical and sensory stability of fried-salted soybeans prepared in different vegetable oils. *J. Am. Oil Chem. Soc.* **2012**, *89*, 1699–1711. [CrossRef]
68. Carpenter, R.P.; Lyon, D.H.; Hasdell, T.A. *Guidelines for Sensory Analysis in Food Product Development and Quality Control*, 2nd ed.; Aspen Publishers, Inc.: Gaithersburg, MD, USA, 2000; pp. 1–11. ISBN 0-8342-1642-6.
69. Moskowitz, H.R.; Beckley, J.H.; Resurreccion, A.V.A. *Sensory and Consumer Research in Food Product Design and Development*; Black Publishing: Ames, IA, USA, 2006; pp. 219–293. ISBN 0-8138-1632-7.
70. ASTM International. *Standard Guide for Sensory Evaluation Methods to Determine the Sensory Shelf Life of Consumer Products*; American Society for Testing of Materials: West Conshohocken, PA, USA, 2005.
71. Riveros, C.G.; Mestrallet, M.G.; Gayol, M.F.; Quiroga, P.R.; Nepote, V.; Grosso, N.R. Effect of storage on chemical and sensory profiles of peanut pastes prepared with high-oleic and normal peanuts. *J. Sci. Food Agric.* **2010**, *90*, 2694–2699. [CrossRef] [PubMed]
72. Riveros, C.G.; Mestrallet, M.G.; Quiroga, P.R.; Nepote, V.; Grosso, N.R. Preserving sensory attributes of roasted peanuts using edible coatings. *Int. J. Food Sci. Technol.* **2013**, *48*, 850–859. [CrossRef]
73. Johnsen, P.B.; Civille, G.V.; Vercellotti, J.R.; Sanders, T.H.; Dus, C.A. Development of a lexicon for the description of peanut flavor. *J. Sens. Stud.* **1988**, *3*, 9–17. [CrossRef]
74. Lee, J.; Chambers, D.H. Flavors of green tea change little during storage. *J. Sens. Stud.* **2010**, *25*, 512–520. [CrossRef]
75. Lee, J.; Chambers, D.H. A lexicon for flavor descriptive analysis of green tea. *J. Sens. Stud.* **2007**, *22*, 256–272. [CrossRef]
76. Poehlman, J.M.; Sleper, D.A. *Breeding Field Crops*, 4th ed.; Iowa State University Press: Ames, IA, USA, 1995; pp. 3–15. ISBN 0-8138-2427-3.
77. Vara-Ubol, S.; Chambers, E., IV; Kongpensook, V.; Oupadissakoon, C.; Yenket, R.; Retiveau, A. Determination of the sensory characteristics of rose apples cultivated in Thailand. *J. Food Sci.* **2006**, *71*, S547–S552. [CrossRef]

© 2019 by the author. Licensee MDPI, Basel, Switzerland. This article is an open access article distributed under the terms and conditions of the Creative Commons Attribution (CC BY) license (http://creativecommons.org/licenses/by/4.0/).

Article

Temporal Drivers of Liking Based on Functional Data Analysis and Non-Additive Models for Multi-Attribute Time-Intensity Data of Fruit Chews

Carla Kuesten [1,*] and Jian Bi [2]

1. Amway, Ada, MI 49355, USA
2. Sensometrics Research and Service, Richmond, VA 23236, USA; bbdjcy@aol.com
* Correspondence: carla.kuesten@amway.com; Tel.: +1-616-787-5673

Received: 30 April 2018; Accepted: 30 May 2018; Published: 3 June 2018

Abstract: Conventional drivers of liking analysis was extended with a time dimension into temporal drivers of liking (TDOL) based on functional data analysis methodology and non-additive models for multiple-attribute time-intensity (MATI) data. The non-additive models, which consider both direct effects and interaction effects of attributes to consumer overall liking, include Choquet integral and fuzzy measure in the multi-criteria decision-making, and linear regression based on variance decomposition. Dynamics of TDOL, i.e., the derivatives of the relative importance functional curves were also explored. Well-established R packages 'fda', 'kappalab' and 'relaimpo' were used in the paper for developing TDOL. Applied use of these methods shows that the relative importance of MATI curves offers insights for understanding the temporal aspects of consumer liking for fruit chews.

Keywords: relative importance of attributes to liking; multicollinearity; temporal drivers of liking (TDOL); functional data analysis; Choquet integral; fuzzy measure; multi-criteria decision-making; Shapley value; interaction indices; LMG statistic; multi-attribute time-intensity (MATI) data; fruit chews

1. Introduction

Drivers of consumer liking is an important topic of sensory and consumer research. Drivers of liking can be defined as the attributes, which have the most important effects on overall liking [1]. Identification of the drivers is to determine relative importance of the explanatory attributes to liking in a model for sensory attributes and liking. However, determining relative importance of attributes to liking is a tricky task when two or more explanatory variables are nearly linearly dependent, i.e., multicollinearity, which are common situations for sensory attributes. There are some advanced mathematical and statistical techniques to measure relative importance of correlated attributes to liking considering both direct effects and interaction effects [2].

Relative importance of attributes to liking may also vary with time. Time-intensity is a dynamic sensory analysis methodology. It measures the intensity of a single sensory perception over time in response to a single exposure to a product or other sensory stimulus. The validity of the time-intensity measurement is based on the fact that sensory perception is a dynamic process and it should be measured dynamically [3]. An observation in conventional sensory analysis methods, such as quantitative descriptive analysis, is a single-point static rating—an integration by the respondent over the course of the sensory experience or in most practicing panels the highest intensity of the attribute at any given point in the experience, while it is a curve in a time-intensity measurement. Hence time-intensity measurement usually provides much more information about sensory properties of products than the conventional sensory analysis methods.

There are some proposed methods for simultaneous evaluations of multiple sensory attributes in time-intensity evaluation, such as the dual-attribute time-intensity method [4], the Temporal

Dominance of Sensations method [5], and the multiple-attribute time-intensity (MATI) method [6]. MATI data in a fruit chews experiment are used in this study.

Temporal drivers of liking (TDOL) is a hot topic in sensory and consumer fields in recent years. See, e.g., [7–10] in recent sensory literature for the topic. The current TDOL methods in the literature mentioned above focus on Temporal Dominance of Sensations data.

1.1. Non-Additive Model for Modeling Panel Descriptive Attributes and Consumer Liking

Multiple regression and other conventional regressions are often used to model consumer liking and sensory attributes. A common assumption for the models (additive models) is that the regression variables are independent of each other. Under this assumption, the models are additive models. This assumption, however, does not hold in many situations. Sensory attributes are usually dependent. It is noted that even if the assumption is not true, the model can be used for prediction of liking without serious difficulty. However, it seriously affects the determination of relative importance of sensory attributes to liking, i.e., determination of drivers of liking [2]. The non-additive models, which consider both direct effects and interaction effects of attributes to consumer liking, are needed to model the correlated sensory attributes and liking.

In the past about two decades, a non-additive model—Choquet integral and fuzzy measure—has been developed and used in the multi-criteria decision-making (also called multi-criteria decision aiding) field and other fields for modeling data in situations where interaction occurs for multiple criteria (attributes). This model is a special regression model. The main advantage compared to the conventional regression models is that this model is able to take into account interactions and dependence between sensory attributes. It can not only provide the predictions of liking, but also give Shapley values, for relative importance of sensory attributes, and interaction measures of each pair of the sensory attributes. See, e.g., [11] for more details about the methodology. Shapley values were proposed by Shapley [12] in game theory. Alfonso [13] introduced the methodology in the sensory field. The R package 'kappalab', which stands for 'laboratory for capacities', can be used for the methodology. See, e.g., [14] for the R package.

1.2. Non-Additive Model for Modeling Both Consumer Attributes and Consumer Liking

There are some different non-additive models for determination of relative importance of correlated attributes. One is the LMG statistic [2,15] and R package 'relaimpo' [16,17]. The LMG statistic is based on linear regression but with variance decomposition, i.e., averaging over orderings of attributes in the model. The LMG statistic was first proposed by Lindeman, Merenda, and Gold [18] and henceforth is named. We can use this model for determination of the relative importance of attributes for both consumer liking and consumer attributes including just-about-right attributes.

1.3. Relative Importance Curves and Temporal Drivers of Liking

Functional data analysis techniques and the R package 'fda' and code 'wfda' can be used on the multiple sets of relative importance measures (Shapley, interaction and LMG statistic) for a series of time points based on the non-additive models to produce the functional curves of relative importance measures.

1.4. Dynamic Aspects of Temporal Drivers of Liking

The dynamic aspects of temporal drivers of liking are the derivatives of the relative importance functional curves, which reflect rates of change of the relative importance measure. Functional data analysis offers direct access to derivatives of functional curves by using a built-in function 'deriv.fd' in the R package 'fda'.

The main objective of this study is to provide a novel TDOL technique which is based on some advanced statistical technologies and is used for MATI data. The TDOL involves, respectively, panel attributes with continuous scales; panel attributes with CATA (check-all-that-apply) scales; and consumer attributes with continuous and just-about-right scales. Functional data analysis [19–21] and non-additive models, i.e., Choquet integral and fuzzy measure in the multi-criteria decision-making [11,13,14], and LMG statistic based on variance decomposition in linear regression [15], are used for the TDOL for MATI data. Well-established R packages 'fda', 'kappalab' and 'relaimpo' for the advanced mathematical and statistical methodologies were used for analysis of the MATI data and TDOL in the paper. The R packages can be downloaded freely from http://cran.r-project.org. Some R codes ('kapf', 'relaf20', 'wfda') used in the paper can be found in the Supplementary File S1in the online version of this paper.

2. Materials and Methods

This research guidance study involves development work on fruit chews using both trained sensory descriptive panel and consumer MATI data collected throughout consumption. The functional ingredients of this fruit chew that provide health benefits pose taste and texture challenges. The fruit components (acerola cherry, baobab, hesperidin, and other citrus bioflavonoids) deliver unpleasant flavors—bitter, astringency, and tannic qualities. The fruit fiber from the acerola cherry creates a fibrous, gritty texture with particulates that require a lot of unpleasant mouth clean-up. Given that the flavor and texture experience during chew-down is particularly relevant for consumer acceptance of this functional food form, MATI was used to help us characterize the sensory flavor and texture profile of the chew.

The objectives of this research were focused on product development and analytical methods. Product understanding and guidance was sought toward optimizing consumer acceptance. Descriptive and consumer results were gathered to understand consumer liking as a function of sensory descriptive and consumer attributes throughout consumption. The analytical research goals were to expand multivariate statistical analyses for MATI data and to extend the conventional drivers of liking analysis with a time dimension to TDOL.

2.1. Descriptive Panel

Nine descriptive panelists were selected that had completed over 100 h of descriptive training using universal scale attribute standards. Universal flavor aromatic reference standards included soda cracker, Mott's applesauce, Minute Maid orange juice, and Welch's grape juice, with intensity ratings of 2, 5, 7 and 10, respectively. The panelists had demonstrated proficiency in terms of lexicon understanding and usage, repeatability, ability to discriminate, scale usage, agreement, and reproducibility prior to participation. Panelists were practiced and proficient in MATI evaluations prior to the study. A pace of 3 s between attributes was used during MATI data collection.

2.2. Consumer Panel

Forty adults who consume chews regularly were selected from among company employees. These consumers ranged in age from 22 to 55 and represented an equal number of males and females. Respondents were screened based on frequency of consumption of chews (product category users), acceptability to the flavors tested, willingness to taste test the chews, no dental issues, non-smokers, normal odor sensitivity (no colds or problems with smelling during the test period), and no known food allergies. All were given a 1-h one-on-one orientation to the MATI program and practiced with a warm-up run. The facilitator of the session made sure that each respondent understood the purpose of the study, demonstrated good eye–hand coordination for data collection, and correctly entered their results. The sample ratings collected from the consumers included: hedonic flavor and texture ratings as well as intensity ratings. Each sample varied in the total time it took respondents to reach complete dissolution (~90 s) and was followed by a 1-min aftertaste. A 2-min break between samples was imposed. Use of in-house panels is routine and well-practiced in industry and considered relevant at early stages of new product development. While a larger sample size would in general be more appropriate for testing with consumers, this smaller sample size of $N = 40$ is justified given that the prototypes being tested are preliminary efforts and exploratory in nature. Product development was looking for direction and guidance at this stage.

2.3. Test Samples

Three test samples were presented; all cherry fruit chews. A retail competitor (Sample C) was used as a warm-up sample. In addition to the functional active ingredients mentioned above, the two test prototypes contained date and raisin paste, pineapple fruit juice concentrate, gum Arabic, fruit solids, sugar cane fiber, magnesium phosphate, humectants, emulsifiers, acids and natural flavors. The prototypes evaluated differed in amount of acerola cherry fiber, described as Sample A (textured prototype) and Sample B (smooth prototype). The cherry fiber content contributed to differences in Smoothness. All samples were square and of similar shape and size (~5.4 gms per piece, 2 pieces per serving). Test samples were produced in the lab, stored at room temperature, and evaluated at one month of age (~30 days old). The retail competitor was purchased in a local grocery and within shelf-life.

2.4. Study Design

2.4.1. Descriptive Panel

The panel evaluated the chews through 6 stages of the 'Texture Breakdown Path' [22] as illustrated in Figure 1, using 2 chews/test sample. The blind-coded prototype samples were randomized and balanced across sessions. The first chew was used to evaluate surface properties and the first bite. A second chew was used through chew-down and the remaining stages. This allowed recording onset and duration for each stage and events such as 'stuck on teeth' and the moment when the chew was completely dissolved. During evaluation of the first chew for surface properties and first bite, panelists were on 'free time' (time used was captured but panelists were allowed to record ratings at their own pace). Initiating chew-down with the second chew re-started the timer thru to swallow, pacing panelists at 3-s per attribute. Swallowing triggered restart of the timer again and began the 60-s aftertaste. The CATA items were checked if presented at an intensity of 3 or higher on a 0–15 point structured line scale, not as dominate sensation as in Temporal Dominance of Sensations. The line scale data was captured in tenths of a unit for a total of 0–150 points.

Stage	Line Scale 0-150 point	CATA						Timer
Surface Properties		Smooth	Rough	Powdery/Chalky	Gritty	Waxy	Oily	
First Bite	Hardness							
	Cohesiveness	Deformable	Springy	Uniform Bite	Dense	Toothpull		FREE
	Moistness of Mass							
	Awareness of Particulates							
Chew Down								START
		Fruit	Sweet	Sour	Bitter	Off-Notes		
Break Down								
		Fruit	Sweet	Sour	Bitter	Off-Notes		
								END
Swallow	Moistness of Mass	Smooth	Rough	Thin	Thick	Throat Irritation		START
	Loose Particles							
Residual (60 seconds)		Astringent	Powdery/Chalky	Gritty	Waxy	Oily		END

Figure 1. Multiple-attribute time-intensity (MATI) descriptive panel scorecard. CATA, Check-All-That-Apply.

2.4.2. Consumer Panel

Consumers evaluated flavor and texture in 2 separate sessions; in addition, 2 separate sessions for each modality were used to apply different scales. For flavor, the following attributes were captured: Overall Liking of Flavor, Fruit Flavor, Sweetness, Sourness and Pleasantness of Aftertaste; for texture, the following attributes: Overall Liking of Texture, Chewiness Stickiness, Smoothness and Pleasantness of Mouthfeel. Ratings were collected as hedonic and intensity line scales or (in separate sessions) as hedonic and just-about-right category scales. Each consumer completed a total of 4 test sessions, each session approximately 20 min in duration—2 flavor and 2 texture sessions completing 3 samples, 1 rep per session. A cross-over design was applied to balance order of modality as well as scale type across the consumers; the prototype samples were randomized and balanced across serving positions.

2.5. MATI Data Analysis

Statistical analyses developed and applied for this work are extensive. Functional data analysis was used to smooth and present the data graphically. Functional data analysis supports analysis of curves and is particularly applicable to analysis of TI data. Using functional data analysis, we predicted the values of the sensory attributes or liking. Two different approaches (depending on the scale type) were applied. For line scales, relative importance (Shapley values) were estimated using Choquet integrals and fuzzy measures in multi-criteria decision-making. This is a special regression model that can take into account interaction and dependence between sensory attributes. For category scales, 9-point hedonic and just-about-right scales, relative importance values (LMG statistics) were estimated using linear regression based on variance decomposition which takes into account correlated attributes and uses averaging over orders [18]. R software was used for the analyses. Further details and examples follow that illustrate the techniques applied to MATI data.

3. Results

Consumers' rated Overall Liking of Texture higher for the smooth prototype and the competitor (Group 1) and lower for the textured prototype (Group 2) (See Figure 2). Figure 3 shows the functional curves for the 2 sensory attributes Hardness and Cohesiveness for the 3 samples. Initially, the competitor is rated hardest, the textured prototype softest and the smooth prototype in-between all peaking around 10-s. The decay curve is relatively steep for all samples, but slightly more gradual for the textured prototype. The samples finish similarly. The competitor and smooth prototype rate similarly and highest over time on Cohesiveness. Though we see more variation with the competitor, the textured prototype is recognized as less cohesive. Figure 4 shows both Moistness of Mass and Awareness of Particles for the 3 samples, again showing more similarity between the competitor and the smooth prototype versus the less moist textured prototype. In addition, we see the obvious issue of particulates for the textured prototype during chew-down. In terms of flavor, Figure 5 shows that consumers indicate their liking decreases over time; they want more Fruit Flavor, Sourness and Sweetness (especially initially and at the end). The proportions of "too strong" are almost zero at any time. The proportions of 'just right' and 'too low' seem to mirror each other—and cross-over/switch around 35 s for intensity of Fruit Flavor, 20 and 80 s for Sweetness and back and forth during 20–80 s for Sourness. Further work is required to optimize the flavor of the smooth prototype.

The results below cover even more depth of understanding for measuring, analyzing and interpreting the presence, duration, change (derivative) and rate of change (second derivative) of attributes in relation to temporal drivers of liking. Understanding how best to leverage these insights and bring them to use within product development optimization will require further study and development efforts.

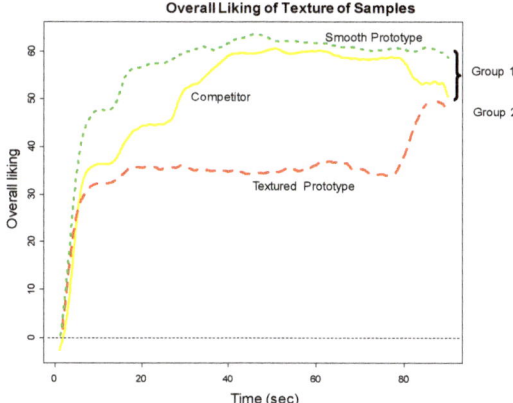

Figure 2. Consumer chew-down Overall Liking of Texture.

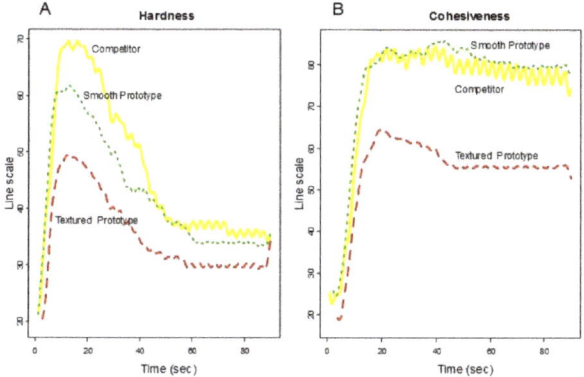

Figure 3. Descriptive panel chew-down (**A**) Hardness and (**B**) Cohesiveness.

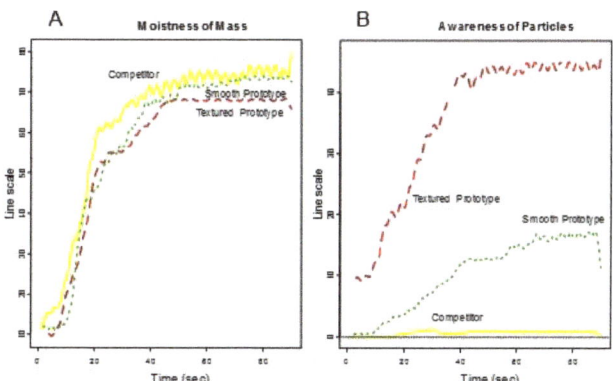

Figure 4. Descriptive panel chew-down (**A**) Moistness of Mass and (**B**) Awareness of Particles.

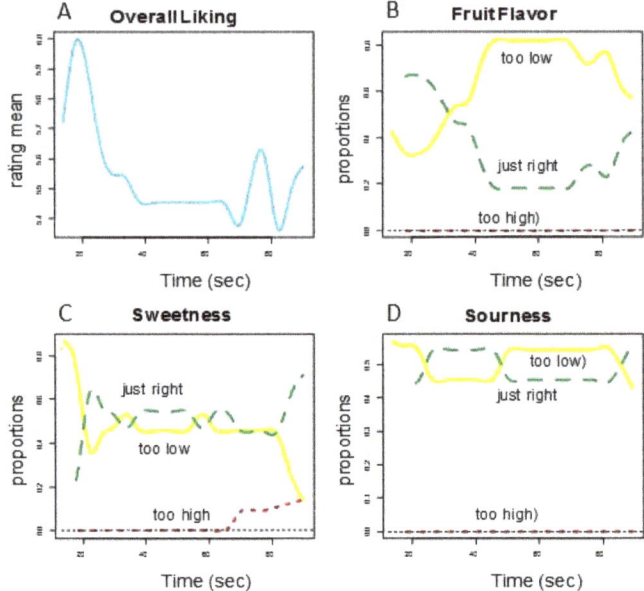

Figure 5. Consumer panel chew-down—smooth prototype (**A**) Overall Liking of Flavor, (**B**) Fruit Flavor, (**C**) Sweetness, and (**D**) Sourness just-about-right scales.

3.1. Non-Additive Model for Modeling Panel Descriptive Attributes and Consumer Liking

The Shapley value, a measure of relative importance, is calculated for each sensory attribute in the R package 'kappalab'. A matrix is given for interaction effects for each pair of the attributes. The sum of the Shapley values for the attributes is one. A positive interaction value in the interaction matrix suggests that the pair of criteria are complementary, while a negative value suggests that they are substitutive.

3.1.1. Numerical Results Output 1 (Panel Attributes with Line Scale)

For the MATI descriptive panel data of the fruit chews experiment, the relative importance values of the 4 panel sensory intensity attributes with line scales (Hardness, Cohesiveness, Moistness of Mass, and Awareness of Particles) to consumer Overall Liking of Texture, and interaction of a pair of the sensory attributes were obtained for each given time point in the range of 0 and 90 s using the R package 'kappalab'. For example, for the 50th second time point, the results are given in Supplementary File S2 using the R code 'kapf'.

In this example, the largest Shapley value is 0.41 for Hardness. The interaction between Hardness and MoistnessOfMass is positive (0.21) and hence the two attributes are complementary, while interaction between Cohesiveness and Awareness of Particles is negative (−0.17), and hence the two attributes are substitutive.

For some time points, e.g., for the 30 time points from 3 to 90 s in a step of 3 s, we estimated the 30 sets (each set is a matrix of 3 × 4) of the values of the 4 panel sensory attributes for 3 samples from the corresponding 3 sets of the functional curves. We also estimated the 30 sets (each set is a vector with a length of 3) of Overall Liking of Texture from the 3 curves of Overall Liking of Texture. With the estimated values for each time point, we calculated the Shapley values of the 4 sensory attributes, which are the relative importance values of the 4 attributes to Overall Liking of Texture at that time point. With the data for 30 time points, we obtained 30 sets of the Shapley values interaction indices for the 4 attributes, which are listed in Tables 1 and 2.

Table 1. Shapley values for each panel descriptive attribute to Overall Liking of Texture varying with time ('rivalue').

Second	Hardness	Cohesiveness	Moistness Of Mass	Awareness Of Particles
3	0.21	0.20	0.18	0.40
6	0.41	0.21	0.17	0.21
9	0.27	0.29	0.20	0.24
12	0.20	0.31	0.20	0.29
15	0.24	0.25	0.25	0.26
18	0.25	0.25	0.25	0.26
21	0.24	0.25	0.24	0.27
24	0.25	0.24	0.24	0.27
27	0.25	0.24	0.24	0.26
30	0.35	0.19	0.19	0.27
33	0.33	0.22	0.20	0.25
36	0.32	0.29	0.20	0.19
39	0.44	0.31	0.14	0.12
42	0.44	0.31	0.14	0.11
45	0.43	0.32	0.14	0.11
48	0.42	0.31	0.15	0.12
51	0.42	0.30	0.16	0.12
54	0.42	0.30	0.16	0.12
57	0.42	0.30	0.16	0.12
60	0.41	0.30	0.16	0.12
63	0.41	0.30	0.16	0.13
66	0.42	0.30	0.16	0.12
69	0.43	0.29	0.16	0.12
72	0.44	0.29	0.16	0.12
75	0.39	0.28	0.17	0.16
78	0.39	0.28	0.17	0.16
81	0.40	0.30	0.17	0.13
84	0.30	0.29	0.21	0.19
87	0.27	0.29	0.24	0.21
90	0.30	0.29	0.21	0.20

Table 2. Interaction indices for each pair of panel descriptive attributes to Overall Liking of Texture varying with time ('intvalue').

Second	"HvC"	"HvM"	"HvA"	"CvM"	"CvA"	"MvA"
3	0.04	−0.03	0.08	−0.02	0.10	0.16
6	−0.06	−0.15	−0.09	−0.04	0	0.04
9	0.13	−0.06	0	−0.1	−0.03	0.03
12	−0.07	0	0.06	−0.08	−0.01	0.06
15	0	0.01	0.02	0	0.01	−0.01
18	0	0	0	0	0.01	0
21	0.01	−0.01	0.02	0	0.01	0.03
24	0.01	−0.01	0.01	0	0.01	0.02
27	0	0.01	0	0	0.01	0.01
30	0.17	0.17	−0.33	0.04	0.04	0.04
33	0.12	0.15	−0.27	0.02	0	0.02
36	0.05	0.12	−0.06	−0.06	−0.08	−0.01
39	0.06	0.24	0.09	−0.01	−0.16	−0.02
42	0.06	0.25	0.08	0	−0.17	−0.03
45	0.05	0.24	0.07	−0.01	-0.18	−0.02
48	0.05	0.22	0.06	0.03	−0.17	−0.03
51	0.05	0.23	0.06	0.03	−0.17	−0.04
54	0.05	0.22	0.06	0.04	−0.16	−0.04
57	0.05	0.22	0.06	0.04	−0.17	−0.04
60	0.05	0.22	0.06	0.04	−0.17	−0.04
63	0.04	0.21	0.06	0.05	−0.17	−0.04
66	0.05	0.22	0.06	0.04	−0.17	−0.04
69	0.06	0.23	0.07	0.04	−0.16	−0.04
72	0.06	0.23	0.07	0.04	−0.16	−0.04
75	0.02	0.18	0.08	0.07	−0.12	−0.01
78	0.02	0.18	0.08	0.07	−0.13	−0.01
81	0.04	0.20	0.05	0.07	−0.16	−0.04
84	−0.02	0.06	0.08	0.05	−0.09	−0.02
87	−0.03	0.02	0.07	0.05	−0.07	−0.03
90	−0.02	0.07	0.07	0.07	−0.09	−0.01

"HvC": Hardness versus Cohesiveness; "HvM": Hardness versus MoistnessOfMass; "HvA": Hardness versus AwarenessOfParticles; "CvM": Cohesiveness versus MoistnessOfMass; "CvA": Cohesiveness versus AwarenessOfParticles; "MvA": MoistnessOfMass versus AwarenessOfParticles.

3.1.2. Numerical Results Output 2 (Panel Attributes with CATA Scale)

In the panel MATI experiment, there is a 'check-all-that-apply' (CATA) question with 5 CATA items (UniformBite, Deformable, Dense, Springy, and Toothpull) in First Bite. We determined the relative importance of each CATA item to consumer Overall Liking of Texture at a specified time point. The data for each item is a proportion. The R package 'kappalb' and code 'kapf' were used. See Supplementary File S3.

For example, for time at 5 s, the relative importance values, i.e., the Shapley values for the 5 CATA items, are obtained below. It suggests that the item UniformBite is the most important among the 5 items to Overall Liking of Texture at 5 s with Shapley value of 0.42. The interaction effect between Deformable and UniformBite is −0.47. The negative interaction suggests that the two attributes are substitutive.

For each time point from 1 to 25 s, we obtained the data, i.e., the proportions of each of the 5 CATA items for the 3 products, and rating means of consumer Overall Liking of Texture at that time. Hence, we derived the 25 sets of Shapley values for the CATA items, which are listed in Table 3.

Table 3. Shapley values for each panel descriptive CATA item to Overall Liking of Texture varying with time ('rivalue2').

Second	Deformable	Springy	UniformBite	Dense	Toothpull
1	0.48	0.12	0.14	0.14	0.11
2	0.28	0.08	0.28	0.28	0.08
3	0.25	0.13	0.10	0.43	0.09
4	0.49	0.12	0.14	0.14	0.11
5	0.26	0.11	0.42	0.12	0.09
6	0.26	0.13	0.42	0.10	0.09
7	0.35	0.09	0.35	0.09	0.13
8	0.38	0.08	0.08	0.08	0.37
9	0.38	0.08	0.08	0.08	0.37
10	0.60	0.10	0.10	0.10	0.10
11	0.60	0.10	0.10	0.10	0.10
12	0.60	0.10	0.10	0.10	0.10
13	0.60	0.10	0.10	0.10	0.10
14	0.60	0.10	0.10	0.10	0.10
15	0.60	0.10	0.10	0.10	0.10
16	0.60	0.10	0.10	0.10	0.10
17	0.60	0.10	0.10	0.10	0.10
18	0.38	0.08	0.37	0.08	0.08
19	0.08	0.08	0.37	0.37	0.08
20	0.08	0.08	0.37	0.37	0.08
21	0.10	0.10	0.6	0.10	0.10
22	0.37	0.08	0.38	0.08	0.08
23	0.28	0.08	0.28	0.28	0.08
24	0.28	0.08	0.28	0.28	0.08
25	0.08	0.08	0.37	0.37	0.08

3.2. Non-Additive Model for Modeling Both Consumer Attributes and Consumer Liking

For each time point, we generated a data matrix with n rows (consumer panelists) and k + 1 columns for Overall Liking and k attributes. We then calculated relative importance of the attributes in terms of the LMG statistic, which is a measure of relative importance as a Shapley value. The LMG statistic is equivalent to the Shapley value in game theory [12]. See e.g., [15] for the relationship between LMG and the Shapley value.

Numerical Results Output 3 (Consumer Just-About-Right Attributes with 3 Levels)

For the consumer MATI data in the fruit chews experiment, we determined the relative importance of 3 consumer just-about-right attributes (Fruit Flavor, Sweetness and Sourness) to Overall Liking of Flavor for Sample B. The just-about-right attributes are treated as factors with 3 levels (−1 = too weak, 0 = just about right, +1 = too strong). For example, for the time point at 14 s, the values of relative importance for the 3 just-about-right attributes Fruit Flavor, Sweetness, and Sourness are 0.27, 0.64 and 0.08, respectively. At time 60 s, the relative importance values are 0.64, 0.12 and 0.24, respectively. At time 90 s, the values are 0.54, 0.33 and 0.24, respectively. Obviously, the relative importance of the attributes to Overall Liking of Flavor varies with time. See Supplementary File S4 for analysis and results.

We calculated the LMG values of the 3 just-about-right attributes at 20 time points (14 to 90 s with a step of 4) for Sample B, which are listed in Table 4. It is noted that using linear regression with variance decomposition, i.e., averaging over orderings, it provides only relative importance values (e.g., LMG) of the attributes to liking, it does not provide the interaction indices as that in the fuzzy measure and Choquet integral in numerical output 1.

Table 4. LMG values for each consumer just-about-right attribute to Overall Liking of Flavor (Sample B) varying with time ('jarLMG').

Second	Fruit Flavor	Sweetness	Sourness
14	0.27	0.64	0.08
18	0.55	0.35	0.11
22	0.43	0.26	0.31
26	0.20	0.19	0.61
30	0.56	0.18	0.26
34	0.55	0.24	0.21
38	0.37	0.11	0.52
42	0.63	0.09	0.28
46	0.38	0.13	0.48
50	0.53	0.16	0.31
54	0.65	0.11	0.25
58	0.64	0.12	0.24
62	0.65	0.15	0.21
66	0.65	0.15	0.21
70	0.51	0.37	0.12
74	0.50	0.44	0.06
78	0.51	0.44	0.05
82	0.38	0.50	0.12
86	0.33	0.54	0.13
90	0.33	0.54	0.13

3.3. Relative Importance Curves and Temporal Drivers of Liking

For multiple sets of relative importance measures (Shapley, interaction and LMG) that we have obtained in Tables 1–4 for a series of time points based on non-additive models, we produced the functional curves of relative importance measures by using functional data analysis techniques and the R package 'fda' and code 'wfda'. The functional objects for the relative importance measures based on the observed data in Tables 1–4 were produced. See Supplementary File S5.

3.3.1. Numerical Results Output 1

For the functional object 'rivaluefd', we generated the relative importance functional curves. See code in Supplementary File S6 used to generate the plot.

Figure 6 gives the Shapley curves of the 4 sensory attributes, which suggest the TDOL. From Figure 6, we see that after about 30 s, the attribute Hardness is the driver of Overall Liking of Texture. In the range from 1 to 30 s, the situation is mixed and unstable.

Figure 7 gives the interaction effect curves for each of the 6 pairs of the 4 sensory attributes based on the functional object 'intvaluefd'. It seems that after about 40 s, the interaction effects between *Hardness* and each of the other attributes, and between Cohesiveness and MoistnessOfMass are consistently positive, i.e., complementary, while Cohesiveness vs. AwarenessOfParticles and MoistnessOfMass vs. AwarenessOfParticles are consistently negative, i.e., substitutive.

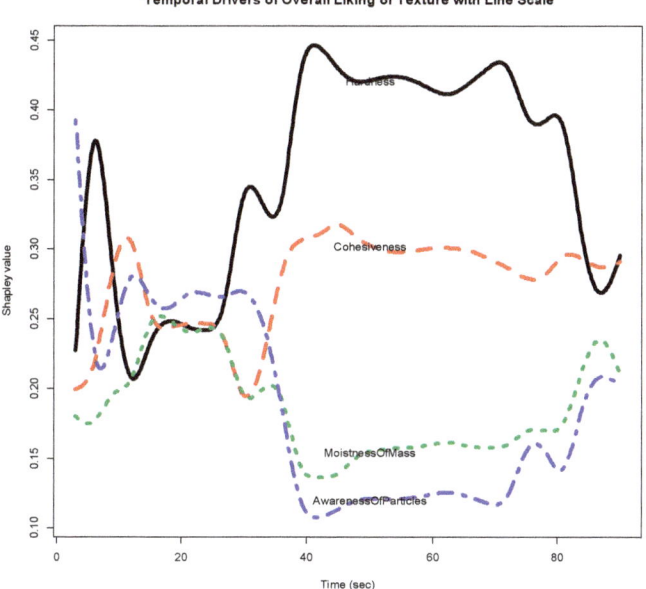

Figure 6. Temporal drivers of Overall Liking of Texture with line scale.

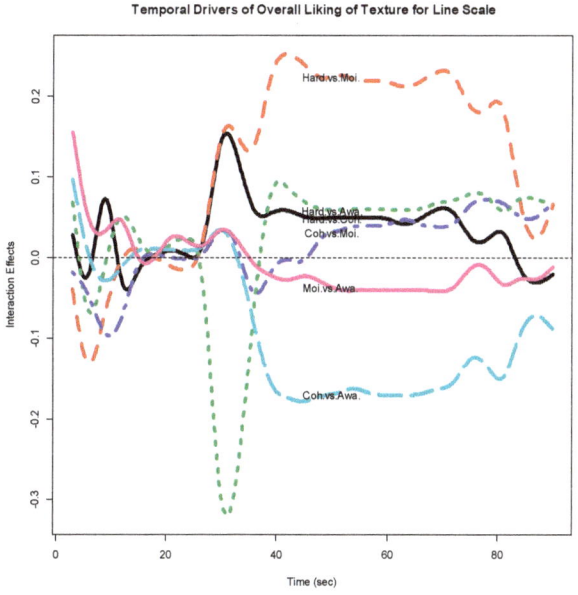

Figure 7. Temporal drivers of Overall Liking of Texture with line scale (interaction).

3.3.2. Numerical Results Output 2

Figure 8 gives the Shapley curves of the 5 CATA items based on the functional object 'rivalue2fd'. The curves reflect temporal relative importance of the CATA items to Overall Liking with time. It should be mentioned that the interaction indices could also be obtained.

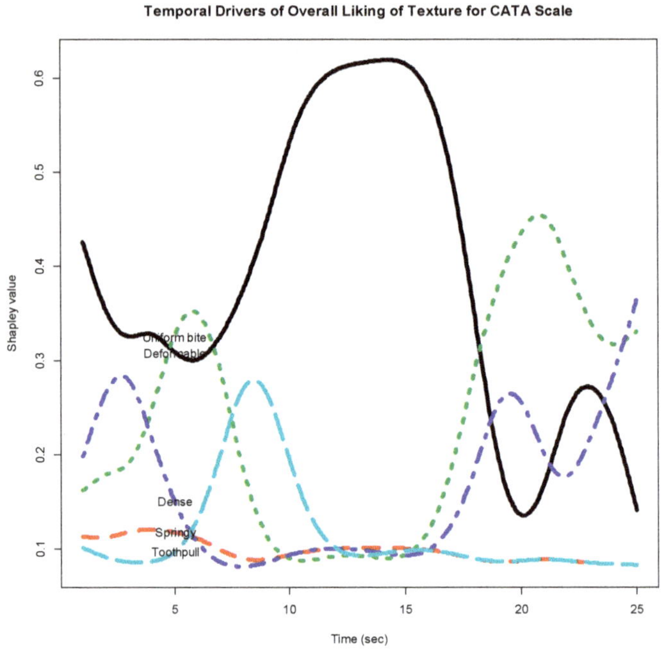

Figure 8. Temporal drivers of Overall Liking of Texture with CATA scale (5 CATA items in the first bite).

3.3.3. Numerical Results Output 3

Figure 9 gives the LMG curves of the 3 just-about-right attributes to Overall Liking of Flavor based on the functional object 'jarRI.fd'. The curves in Figure 9 show possible temporal drivers of liking analysis for just-about-right attributes for both consumer liking and consumer just-about-right attributes. It suggests that the relative importance of the 3 just-about-right attributes to Overall Liking of Flavor varies with time. In the beginning and the end (after about 80 s), Sweetness seems the most important effect on liking within the 3 attributes, while during most of the time (about 30 to 80 s), Fruit Flavor is the driver of Overall Liking.

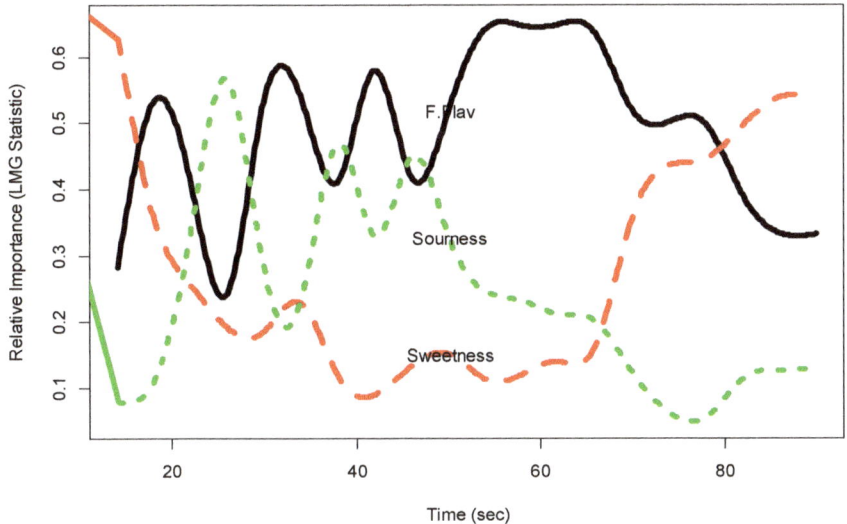

Figure 9. Temporal drivers of Overall Liking of Texture with just-about-right scale.

3.4. Dynamic Aspects of Temporal Drivers of Liking

The first and second derivatives of the functional objects in Figures 6–9 reflect rates of change of the relative importance measures. These are obtained as shown in Supplementary File S7.

We produced the charts of the first and second derivatives of the function object, e.g., 'rivaluefd'; it is shown in Figure 10. See Supplementary File S7.

Figure 10 roughly shows the first and second derivatives of the temporal DOL curves in Figure 6 for the numerical outputs 1–2 in this paper.

The key of this exploration is about the practical meaning and interpretation for derivatives of the relative importance functional curves. The first and second derivatives of the TDOL curves show the velocities and accelerations of the relative importance measures. The values of zero in the first derivative of the relative importance measure correspond to local maximum or minimum values of the relative importance measures. If the absolute values of the first and second derivatives for an attribute are close to zero, it suggests that the relative importance of the attribute to liking is stable.

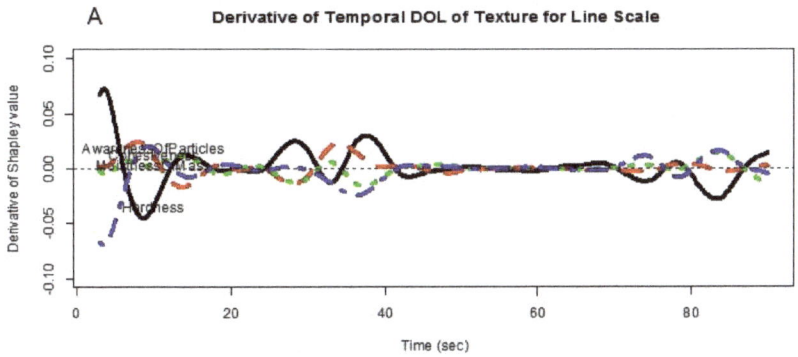

Figure 10. *Cont.*

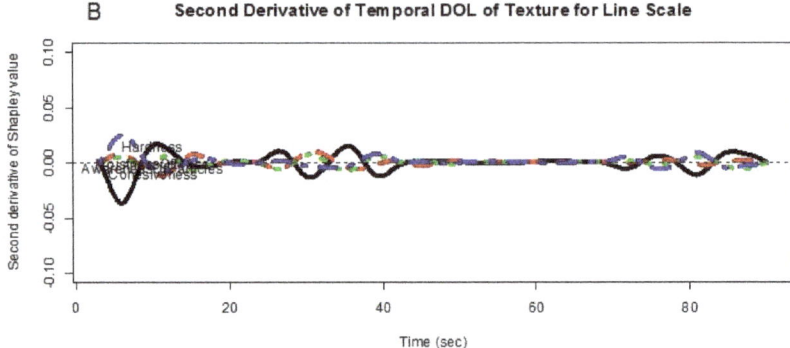

Figure 10. Derivatives of temporal drivers of Overall Liking of Texture with line scale. (**A**) First Derivative and (**B**) Second Derivative.

4. Discussion

This research was pursued due to a concern about consumer texture acceptability; the fruit chews under investigation provided an undesirable lack of cohesiveness and moistness with high awareness of particulates during chew-down due to the fiber content. Research efforts were undertaken to determine if modifications could be made to improve the texture experience for consumers. Evidence of the improvement is highlighted with the use of MATI TDOL, wherein we see that the initial prototype (textured prototype) demonstrates less Cohesiveness and Moistness during chew-down compared to a modified prototype (smooth prototype) which more closely approximates the competitor for Overall Liking of Texture.

4.1. MATI Product Research

Practical, applied learnings for product development and optimization have been gleaned from this work. This study has shown that MATI offers the ability to relate sensory and consumer data on a time-dependent basis. Conventional drivers of liking DOL analyses can be extended with a time dimension into TDOL. MATI is a natural extension of traditional time-intensity with one attribute. This method serves as a rapid 'real-time' assessment tool for product screening and optimization. As shown here, it was applied successfully for gathering information throughout the Texture Breakdown Path wherein consumption events were logged. Training and practice are required for reliable results; an easy user-interface is important. MATI can serve to aid and enhance panel training experiences. In regard to consumer research with MATI, care should be taken to explain the method thoroughly. How MATI programming is structured for data collection impacts the results. Results may vary from conventional methods.

4.2. MATI Data Analytics

Relative importance of attributes to consumer liking may vary with time. The curves of relative importance of attributes reflect the variations, and hence the conventional drivers of liking were extended into TDOL based on functional data analysis and non-additive models, which consider both direct and indirect effects of an attribute to consumer liking.

There are different non-additive models for determination of relative importance of attributes to liking. The Choquet integral and fuzzy measure in the multi-criteria decision-making are suitable to determination of panel descriptive attributes to *Overall Liking*, in which the number of samples are usually small. The linear regression based on variance decomposition is suitable to determination of both consumer attributes and consumer liking, i.e., original consumer data, because in this method, the number of consumer panelists (rows in the data file) should be larger, at least larger than the number of attributes (columns of the data file).

The dynamic aspects of relative importance curves of attributes may provide additional insights for TDOL. We found at which time points the relative importance of an attribute to liking tends to change, i.e., rates of change (first derivative) and acceleration of change (second derivative). Functional data analysis is a powerful tool for exploring the dynamic aspect of the TDOL.

4.3. Future Research

Sensory and consumer temporal methods have grown in recent years. Future efforts to continue this evolution will undoubtedly involve ongoing comparison of simultaneous time-intensity results with conventional descriptive panel, discrete time-intensity, and consumer data, enhanced experiences for MATI data collection, investigation into variations on protocols, instructions, warm-up, conditioning or priming of initial responses, examination of the impact of interface and design on response ratings, as well as application and extension of MATI to other product categories. Furthering the understanding of consumer temporal multivariate data through analytics and how best to leverage the dynamic aspects of TDOL is on the horizon.

5. Conclusions

The objectives of this research were satisfied through use of MATI with descriptive and consumer data of fruit chews for product research guidance. The work also accomplished the research goals of successfully expanding multivariate statistical analyses for MATI data and extending the conventional drivers of liking analysis with a time dimension to TDOL.

Supplementary Materials: The following are available online at http://www.mdpi.com/2304-8158/7/6/84/s1. Supplementary File S1: R Codes, Supplementary File S2: Numerical Results Output 1 (panel attributes with line scale), Supplementary File S3: Numerical Results Output 2 (panel attributes with CATA scale), Supplementary File S4: Numerical Results Output 3 (Consumer JAR Attributes with 3 Levels), Supplementary File S5: Relative importance curves and temporal drivers of liking, Supplementary File S6: Numerical Results Output 1 (Shapley values plot), Supplementary File S7: Dynamic aspects of temporal drivers of liking.

Author Contributions: C.K. conceived of and designed the work; J.B. analyzed the data, C.K. and J.B. wrote the article; C.K. and J.B. revised and proofread the article.

Funding: This research received no external funding.

Acknowledgments: The authors would like to acknowledge the following individuals for their contribution to this research: Gene Maly, R&D Formulator; Troy Nietling, R&D certified Food Scientist; Jenny Wu, Laura Malnor, Don Williams and Yev Gevorkyan, Consumer Product Research Sensory Scientists and Technicians, Noe Galvan, Technical Regulatory and Safety Scientist and the descriptive and consumer panelists who contributed to the studies.

Conflicts of Interest: The authors declare no conflict of interest.

References

1. Moskowitz, H.R. Margarine: The drivers of liking and image. *J. Sens. Stud.* **2001**, *16*, 53–72. [CrossRef]
2. Bi, J. A review of statistical methods for determination of relative importance of correlated predictors and identification of drivers of consumer liking. *J. Sens. Stud.* **2012**, *27*, 87–101. [CrossRef]
3. Dijksterhuis, G.B.; Piggott, J.R. Dynamic methods of sensory analysis. *Trends Food Sci. Technol.* **2001**, *11*, 284–290. [CrossRef]
4. Duizer, L.M.; Bloom, K.; Findlay, C.J. Dual-attribute time-intensity sensory evaluation: A new method for temporal measurement of sensory perceptions. *Food Qual. Preference* **1997**, *8*, 261–269. [CrossRef]
5. Pineau, N.; Schlich, P.; Cordelle, S.; Schlich, P. Temporal Dominance of Sensation: A new technique to record several sensory attributes simultaneously over time. In Proceedings of the 5th Pangborn Symposium, Boston, MA, USA, 20–24 July 2003; p. 121.
6. Kuesten, C.; Bi, J.; Feng, Y.-H. Exploring taffy product consumption experiences using a multi-attribute time-intensity (MATI) method. *Food Qual. Preference* **2013**, *30*, 260–273. [CrossRef]

7. Thomas, A.; Vaisalli, M.; Cordelle, S.; Schlich, P. Temporal drivers of liking. *Food Qual. Preference* **2015**, *40*, 365–373. [CrossRef]
8. Meyners, M. Testing for differences between impact of attributes in penalty-lift analysis. *Food Qual. Preference* **2016**, *47*, 29–33. [CrossRef]
9. Carr, B.T.; Lesniauskas, R.O. Analysis of variance for identifying temporal drivers of liking. *Food Qual. Preference* **2016**, *47*, 97–100. [CrossRef]
10. Castura, J.C.; Li, M. Using TDS dyads and other dominance sequences to characterize products and investigate liking changes. *Food Qual. Preference* **2016**, *47*, 109–121. [CrossRef]
11. Grabisch, M.; Labreuche, C. A decade of application of the Choquet and Sugeno integrals in multi-criteria decision aid. *Ann. Oper. Res.* **2010**, *175*, 247–286. [CrossRef]
12. Roth, A.E. *A value for n-person games. The Shapley Value: Essays in Honor of Lloyd S. Shapley*; Cambridge University Press: Cambridge, UK, 1988.
13. Alfonso, L. A technical note on the use of Choquet integral to analyze consumer preferences: Application to meat consumption. *J. Sens. Stud.* **2013**, *28*, 467–473. [CrossRef]
14. Grabisch, M.; Kojadinovic, I.; Meyer, P. A review of capacity identification methods for Choquet integral based multi-attribute utility theory—Application of the Kappalab R package. *Eur. J. Oper. Res.* **2008**, *186*, 766–785. [CrossRef]
15. Grömping, U. Relative importance in linear regression based on variance decomposition. *Am. Stat.* **2007**, *61*, 139–147. [CrossRef]
16. Grömping, U. Relative importance for linear regression in R: The package relaimpo. *J. Stat. Softw.* **2006**, *17*, 1–27. [CrossRef]
17. *R Package 'relaimpo' Manual*; CRAN, 2010.
18. Lindeman, R.H.; Merenda, P.F.; Gold, R.Z. *Introduction to Bivariate and Multivariate Analysis*; Scott Foresman: Glenview, IL, USA, 1980.
19. Ramsay, J.O.; Silverman, B.W. *Functional Data Analysis*, 2nd ed.; Springer: New York, NY, USA, 2005.
20. Ramsay, J.O.; Hooker, G.; Graves, S. *Functional Data Analysis with R and MATLAB*; Springer: New York, NY, USA, 2009.
21. Bi, J.; Kuesten, C. Using functional data analysis (FDA) methodology and the R package 'fda' for sensory time-intensity evaluation. *J. Sens. Stud.* **2013**, *28*, 474–482. [CrossRef]
22. Hutchings, J.B.; Lillford, P.J. The perception of food texture-The philosophy of the breakdown path. *J. Texture Stud.* **1988**, *19*, 103–115. Available online: http://cran.r-project.org/web/packages/relaimpo/relaimpo.pdf (accessed on 1 June 2018). [CrossRef]

© 2018 by the authors. Licensee MDPI, Basel, Switzerland. This article is an open access article distributed under the terms and conditions of the Creative Commons Attribution (CC BY) license (http://creativecommons.org/licenses/by/4.0/).

Article

Novel Modelling Approaches to Characterize and Quantify Carryover Effects on Sensory Acceptability

Damir Dennis Torrico [1,2,*], **Wannita Jirangrat** [2], **Jing Wang** [3], **Penkwan Chompreeda** [4], **Sujinda Sriwattana** [5] **and Witoon Prinyawiwatkul** [2]

1. School of Agriculture and Food, Faculty of Veterinary and Agricultural Sciences, The University of Melbourne, Parkville, VIC 3010, Australia
2. School of Nutrition and Food Sciences, Louisiana State University Agricultural Center, Baton Rouge, LA 70803, USA; Wannita.Jirangrat@thaiunion.com (W.J.); wprinya@lsu.edu (W.P.)
3. College of Nursing and Health Innovation, University of Texas, Arlington, TX 76019, USA; jing.wang@uta.edu
4. Department of Product Development, Faculty of Agro-Industry, Kasetsart University, Bangkok 10900, Thailand; fagipkc@ku.ac.th
5. Sensory Evaluation and Consumer Testing Unit, Faculty of Agro-Industry, Chiang Mai University, Chiang Mai 50100, Thailand; sujinda.s@cmu.ac.th
* Correspondence: damir.torrico@unimelb.edu.au; Tel.: +61-3-8344-9698

Received: 12 October 2018; Accepted: 6 November 2018; Published: 8 November 2018

Abstract: Sensory biases caused by the residual sensations of previously served samples are known as carryover effects (COE). Contrast and convergence effects are the two possible outcomes of carryover. COE can lead to misinterpretations of acceptability, due to the presence of intrinsic psychological/physiological biases. COE on sensory acceptability (hedonic liking) were characterized and quantified using mixed and nonlinear models. $N = 540$ subjects evaluated grape juice samples of different acceptability qualities (A = good, B = medium, C = poor) for the liking of color (C), taste (T), and overall (OL). Three models were used to quantify COE: (1) COE as an interaction effect; (2) COE as a residual effect; (3) COE proportional to the treatment effect. For (1), COE was stronger for C than T and OL, although COE was minimal. For (2), C showed higher estimates (-0.15 to $+0.10$) of COE than did T and OL (-0.09 to $+0.07$). COE mainly took the form of convergence. For (3), the absolute proportionality parameter estimate (λ) was higher for C than for T and OL (-0.155 vs. -0.004 to -0.039), which represented -15.46% of its direct treatment effect. Model (3) showed a significant COE for C. COE cannot be ignored as they may lead to the misinterpretation of sensory acceptability results.

Keywords: carryover effects; sensory acceptability; sensory bias; mixed models; nonlinear models

1. Introduction

The sensory biases caused by the residual sensations of previously served samples are known as carryover effects [1–3]. Contrast and convergence effects are the two possible outcomes of carryover effects. The contrast effect is defined as the increased perceived difference or discrepancy among products in a sample set. Conversely, the convergence effect is related to the increased perceived similarity among products [2,4–7]. The contrast and convergence effects are hypothesized to affect hedonic results differently. Acceptability ratings tend to decrease when a poor-quality product is preceded by a good-quality product (contrast effect). Contrariwise, acceptability ratings for a good-quality product tend to decrease when it is preceded by another good-quality product (convergence effect) [3]. This problem is commonly encountered in crossover trials in which subjects receive a sequence of different sample sets in multiple products assessments. The effect of the first sample usually carries over to the second sample, and this process is repeated for any other subsequent sample in the trial set. In crossover experiments, the carryover effect can lead to misinterpretations when differences among products can increase or decrease inappropriately, due to the presence of intrinsic faults in the experimental design [8]. Besides, carryover can also affect the development of new products, where multiple samples are tested disregarding the potential biases.

The positional (related to the order in which samples are presented) and carryover effects may affect the sensory acceptability ratings of products by confounding the estimated parameters of the sample/treatment effect (the unbiased sensory difference among the tested products) with the intrinsic sample order effect (the psychological bias originated from the experimental procedure). In sensory evaluation, several techniques have been proposed to minimize carryover effects during tasting, including extending washout periods and using different palate cleansers [9,10]. Carrot sticks and sparkling mineral water [11], bread and mineral water [12], Melba toast and neutral water [13], unsalted cracker and soda water [14], apple slices and carbon filtered water [15] are some of the cleansers commonly used in sensory tests. In terms of the experimental plan, Jirangrat et al. [8] demonstrated that the split-plot design is a suitable experimental procedure for reducing order effect biases in multiple products testing. Split-plot designs can achieve outcomes that are less susceptible to bias by extracting a larger portion of the explained variance from the error (unexplained variance). However, the intrinsic carryover effects may affect the scores of sensory attributes in consumer tests even with the use of proper experimental designs (randomized complete block design, and/or split-plot design) and protocols for testing (such as providing sufficient time in between samples). In a previous study, the serving order of red wines with different alcohol concentrations affected the sensory perceptions of panelists [16]. Currently, there are limited published works [2,3,17] regarding the identification and quantification of carryover effects on acceptability tests using untrained naïve consumers (whom, due to a lack of training in sensory practices, can be susceptible to different product and testing biases). Besides, the majority of the research has been conducted with trained assessors (10 to 15 panelists), whom may show less psychological biases when assessing foods or beverages [18] since they have acquired certain familiarity levels with the sensory testing procedures. Therefore, quantifying the presence of carryover effects on untrained consumer panelists is still unclear.

For acceptability tests, hedonic liking scores are used to represent the level of satisfaction of consumers toward samples or prototypes. Analysis of variance (ANOVA) has been extensively used to test whether hedonic means from different samples are significantly different, regardless of the presence of order effects or any other confounding bias. ANOVA is conducted under the assumption that all the responses follow a normal distribution, where a minimum or no confounding effect is involved. However, such an assumption is, in most cases, unfulfilled, since psychological and physiological biases generally affect the assessment of products in a sample set. Our previous study [8] identified and quantified the contrast effects of samples in hedonic ratings using ANOVA and baseline logistic regression. However, knowledge about the contrast and convergence effects on acceptability tests is still very limited. In the present study, several alternative statistical models were further investigated, including generalized linear mixed models and nonlinear mixed models.

For mixed effects models, two types of coefficients are estimated, including fixed (a characteristic of the entire population) and random (a characteristic of individual experimental or observational units) effects. Mixed models can fit the data with correlations where the response is not necessarily normally distributed. These correlations can arise from repeated observations/measurements of the same sampling unit. In consumer tests, a set of samples is assessed, in which the responses (scores) are correlated within each assessor. Thus, mixed models are appropriate for analyzing consumers responses within a given experimental design. On the other hand, nonlinear models can be also used to predict responses that do not necessarily follow a linear function. A nonlinear model is an extension of a generalized nonlinear mixed model in which the conditional mean is a nonlinear function that is added to an inverse link of the linear predictors. Lindstrom & Bates [19] and Davidian & Giltinan [20] proposed the use of nonlinear mixed models for repeated measurements. As stated earlier, consumers tests can be considered as an example of repeated measurements, as individuals/panelists (a random sample from the population of interest) score repeatedly under different experimental conditions. By definition, both models (mixed and nonlinear) can be used to investigating the contrast and convergence effects in consumers tests. Thus, the objective of this study was to characterize and quantify carryover effects (due to the psychological bias of consumers that consists in contrasting two products) on sensory acceptability scores using generalized linear mixed models and nonlinear mixed models.

2. Materials and Methods

2.1. Materials

Grape juice was selected as the product model, due to its simplicity and commonality within the USA. Fruit juices, in general, are simpler food/beverage models compared to gels, solids, or semi-solid systems, in terms of sensory perception [21]. The liquid model was selected to avoid the presence of extra physiological biases generated from the mastication and swallowing of the samples. A prescreening of 10 local (Baton Rouge, LA, USA) commercial grape juice samples was performed with 30 panelists who regularly consumed grape juice (at least one or two times a week) prior to the actual experiment. The pre-selected criteria were based on five sensory attributes, including purple color, liquid transparency, grape flavor, sweetness, and sourness. The acceptability test (overall liking using a 9-point hedonic scale) was carried out for each grape juice sample. The aim was to classify the level of liking for each product. The three products, having preliminary (N = 30) liking scores of 8–9 (good), 6–7 (medium), and 4–5 (poor) were selected for this study. To avoid conflicts of interest, the names of these products were not reported.

The consumer evaluation was conducted in the Sensory Analysis Laboratory, School of Nutrition and Food Sciences, Louisiana State University, Agricultural Center, Baton Rouge, LA, USA. A large focus-group type room equipped with multiple tables was used to conduct the sensory test. The test room (temperature 25 ± 2 °C) was illuminated with cool, natural, fluorescent lights. A total of N = 540 consumers (N = 269 females and N = 271 males; >18 years) from a pool of faculty, staff, and students from Louisiana State University were recruited and pre-screened using the following criteria: (1) regular consumers of grape juice based on self-reported responses, and (2) not having taste/smell disorders and/or kidney/liver problems. The use of human subjects in this research was approved by the Louisiana State University Agricultural Center Institutional Review Board (IRB# HE11-29). Consumers were briefed about the questions, particularly the sensory attributes and their meanings, and sample handling during the evaluation. Products were poured into Propak™ Soufflé clear lidded plastic cups (60 mL) (Independent Marketing Alliance, Houston, TX, USA). Each cup was half filled (30 mL) and labeled with 3-digit random codes (generated from the Random Orders of the Digits table [22]).

A balanced crossover design with three treatments and three periods (positions) was carried out in this experiment. This design was uniform within the sequences in the sense that each treatment appeared the same number of times within each sequence. This design was also uniform within periods, meaning that each treatment appeared the same number of times within each period. The design is balanced in the sense that each treatment preceded every other treatment the same number of times. Table 1 shows the 3 × 3 crossover design. Table 2 lists the input design effects and their classification. Three commercial grape juices were classified into three quality categories (good, medium, and poor) according to preliminary liking scores as shown in Table 2. Participants (N = 540) were asked to rate their liking according to two sensory attributes (color and taste), and to rate the overall liking of the grape juice samples using a 9-point hedonic categorical scale (a scale with number and definition) where 1 = extremely dislike, 5 = neither like nor dislike and 9 = extremely like [23]. Three main carryover effects (n = 3) were estimated in the experimental design, due to the residual effects of their corresponding juice treatments (good, medium, and poor quality).

Table 1. A three-period (position) crossover design.

	Subject	Position		
		1	2	3
Sequence ABC	1	$y_{1,111}$ (A)	$y_{1,121}$ (B)	$y_{1,131}$ (C)
	2	$y_{2,111}$ (A)	$y_{2,121}$ (B)	$y_{2,131}$ (C)
	.			
	90	$y_{90,111}$ (A)	$y_{90,121}$ (B)	$y_{90,131}$ (C)
Sequence BCA	1	$y_{1,112}$ (B)	$y_{1,122}$ (C)	$y_{1,132}$ (A)
	2	$y_{2,112}$ (B)	$y_{2,122}$ (C)	$y_{2,132}$ (A)
	.			
	90	$y_{90,112}$ (B)	$y_{90,122}$ (C)	$y_{90,132}$ (A)
Sequence CAB	1			
	.			
	90			
Sequence CBA	1			
	.			
	90			
Sequence ACB	1			
	.			
	90			
Sequence BAC	1			
	.			
	90			

Table 2. Description of the input design effect for the grape juice consumer test.

Input Design Effect	Description	Classification
Treatments/Samples (3)	A (good), B (medium), C (poor) [1]	Fixed
Sequences (6)	ABC, ACB, BAC, BCA, CAB, CBA	Fixed
Sample position (3)	1 (left), 2 (center), 3 (right)	Fixed
Panelists (540)	90 panelists per sequence	Random
Carryover effects (3)	Carryover effect of each treatment	Fixed

[1] Product quality from a hedonic classification: liking scores of 8–9 (good), 6–7 (medium) and 4–5 (poor).

2.2. Statistical Experiment and Analysis

Three grape juice products were served to each participant using one of the six random serving orders (ABC, ACB, BAC, BCA, or CAB, CBA) so that each order was assigned to 90 participants (6 × 90 = 540 in total); in other words, each sequence order (ABC, ACB, BAC, BCA, CAB, or CBA) was assessed 90 times. All products were tasted in a blinded and unbranded manner. To reduce the presentation protocol errors, each participant was exposed to all products (grape juices) at the same time [24].

Water and unsalted crackers were served as palate neutralizers during the experiment. Re-tasting of products was allowed to refresh participants' memory only if needed [25]. After tasting, each participant was asked to rate three sensory aspects (color, taste and overall liking) of each product. All hedonic data were analyzed at $\alpha = 0.05$ using the SAS software 9.4 (SAS Institute Inc. Cary, NC, USA). For fitting the data, three models were considered to identify and quantify the carryover effects as shown below:

Model 1:

$$y_{ijk} = \mu + \alpha_j + \rho_{i(k)} + \tau_{d(j,k)} + \beta_k + \phi_{jk} + \varepsilon_{ijk} \quad (1)$$
$$(i = 1 \ldots 90, j = 1, 2, 3 \text{ and } k = 1, \ldots, 6)$$

where y_{ijk} is the response due to participant i, position j, and sequence k. μ is the overall mean. α_j is the fixed effect due to position j, subject to $\Sigma \alpha_j = 0$. $\rho_{i(k)}$ is the random effect due to subject i being nested within sequence k, assumed to follow a normal distribution with mean zero and a constant variance. $\tau_{d(j,k)}$ is the fixed, direct effect of treatment (grape juice) that is assigned to the jth position and kth sequence, subject to $\Sigma \tau_{d(j,k)} = 0$. β_k is the fixed effect due to sequence k, subject to $\Sigma \beta_k = 0$. ϕ_{jk} is the fixed effect (carryover effect) or the interaction effect $\alpha_j \tau_{d(j,k)}$ between treatment $\tau_{d(j,k)}$ and position k, subject to $\Sigma \phi_{jk} = 0$. ε_{ijk} are independent random errors that are assumed to follow a normal distribution with mean zero and a constant variance. ε_{ijk} and $\rho_{i(k)}$ are independent.

To quantify the magnitude or size and direction of carryover effects in model (1), the carryover effect ϕ_{jk} was replaced with $\gamma_{d(j-1,k)}$, the fixed effect of the first-order carryover or residual effect from the treatment assigned to the $(j-1)$th position of the kth sequence [26].

Model 2:

$$y_{ijk} = \mu + \alpha_j + \rho_{i(k)} + \tau_{d(j,k)} + \beta_k + \gamma_{d(j-1,k)} + \varepsilon_{ijk} \quad (2)$$
$$(i = 1, \ldots, 90, j = 1, 2, 3 \text{ and } k = 1, \ldots, 6)$$

where $\gamma_{d(j-1,k)}$ was the fixed effect of the first-order carryover or residual effect from the treatment assigned to the $(j-1)$th position of the kth sequence, subject to $\Sigma \gamma_{d(j-1,k)} = 0$. Note that there was no carryover effect of the treatment assigned to the first position in each sequence, i.e., $\gamma_{d(0,k)} = 0$.

Ferris et al. [2] suggested a proportional relationship between the treatment carryover effect and its direct effect, such that $\gamma_{d(j-1,k)} = \lambda \tau_{d(j,k)}$, where $|\lambda| < 1$ in general, i.e., model (2) becomes model (3).

Model 3:

$$y_{ijk} = \mu + \alpha_j + \rho_{i(k)} + \tau_{d(j,k)} + \beta_k + \lambda \tau_{d(j,k)} + \varepsilon_{ijk} \quad (3)$$
$$(i = 1, \ldots, 90, j = 1, 2, 3 \text{ and } k = 1, \ldots, 6)$$

The sign of λ indicates the form of carryover. When $\lambda > 0$, there is an assimilation of the previous treatment, and when $\lambda < 0$, carryover takes the form of a contrast effect.

3. Results and Discussion

Table 3 presented the type I test of fixed effects estimated from model (1) for three sensory attributes (color, taste, and overall liking). For all sensory attributes, the treatment effect was significant ($p < 0.0001$), indicating that the grape juice samples (A, B, and C) were different in their hedonic scores. As expected, grape juice A had a higher liking score for color, taste, and overall liking (7.36–7.74) than B (6.37–6.71) and C (3.50–3.90) (Table 4). In addition, the position effect (left = 1, center = 2, and right = 3) was significant for taste ($p = 0.02$), but not significant for color ($p = 0.31$) and overall liking ($p = 0.24$). For all attributes (color, taste, and overall liking), the sequence and carryover effects were not significant ($p \geq 0.05$). However, the carryover effect was stronger for color (F-value = 1.94) than for taste (F-value = 0.63) and overall liking (F-value = 0.59).

Table 3. Type I test of fixed effects evaluated on three attributes (color, taste, and overall liking) from model (1).

Attribute	Effect	Num DF [1]	Den DF [1]	F-Value [2]	Pr > F [2]
Color	Sequence	5	534	2.00	0.0778
	Position	2	1074	1.18	0.3072
	Treatment	2	1074	1180.95	*<0.0001*
	Carryover	2	1074	1.94	0.1440
Taste	Sequence	5	534	1.60	0.1595
	Position	2	1074	3.91	*0.0204*
	Treatment	2	1074	558.52	*<0.0001*
	Carryover	2	1074	0.63	0.5333
Overall Liking	Sequence	5	534	1.15	0.3328
	Position	2	1074	1.43	0.2388
	Treatment	2	1074	706.40	*<0.0001*
	Carryover	2	1074	0.59	0.5549

[1] Num DF = Degrees of freedom of numerator, Den DF = Degrees of freedom of denominator. [2] F-value = Mean square/Mean square error. F-value under the hypothesis of Ho: Effect = 0. Effects were considered significant when the probability Pr > F was less than 0.05 (Bolded and italicized probabilities).

Table 4. Mean values [1] for the sensory acceptability scores of the grape juice samples from model (1).

Samples [2]	Color	Taste	Overall Liking
A	7.36 ± 1.49 [a]	7.74 ± 1.30 [a]	7.40 ± 1.45 [a]
B	6.71 ± 1.32 [b]	6.37 ± 1.75 [b]	6.37 ± 1.60 [b]
C	3.50 ± 1.75 [c]	3.90 ± 2.08 [c]	3.74 ± 1.92 [c]

[1] Data are represented as mean and standard deviation values (N = 540). Liking scores were based on a 9-point hedonic scale (1 = dislike extremely, 9 = like extremely). [2] Samples descriptions are shown in Table 2. [a–c] mean values with different letters within the same column for each parameter are significantly different ($p < 0.05$).

In consumer sensory trials, the first order and carryover effects can be minimized by balancing the order of sample presentations when each treatment occurs an equal number of times in each position [27]. Suitable experimental designs such as cross-over (change-over) and Williams designs can be used for minimizing the effects of carryover between samples [28]. In the present study, a Balanced Randomized Block Design (BRBD) was carried out, in which each sample occurs an equal number of times in each position of the trial. Moreover, the 3 × 3 crossover design ensured that all possible adjacent pair of treatments (AB, BA; AC, CA; BC, CB) occurred an equal number of times. The last position (right) had a significantly ($p < 0.05$) lower liking score compared to the first (left) and center positions (5.7 vs. 5.9–6.0 scores for taste; data not shown). The order of tasting can introduce a significant bias in sensory evaluations [28]. In the case of taste, the lower scores reported for the last position (right) in the present study may be partially explained by a common psychological bias of consumers, in which the first sample often receives the highest score in a sequence of samples [29]. However, order and carryover effects are dependent on the sensory context including the nature of the product, attribute selection, and the training level of panelists [30].

Model (2) had the advantage of measuring the magnitude and direction of the treatment carryover effect. In this model, the liking of color showed higher absolute mean estimates of the carryover effect size (−0.15 to +0.10) than did the liking of taste (−0.10 to +0.07) and the overall liking (−0.09 to +0.07) (Table 5), although none of them were significant. The carryover effect represented the size of the bias that can affect the score of the following sample in the sequence. For instance, if the size of the effect was +0.10, this means that the following sample would be positively biased by 0.10 units in the hedonic score. Furthermore, on the 9-point hedonic scale, this carryover effect size +0.10 accounted for (0.1/8) × 100% = 1.25% (considering that the length of the 9-point hedonic scale is 8 units) of its direct treatment effect. In fact, the liking of color had the higher carryover effect bias (−1.90% to 1.29%) than did the liking of taste (−1.26% to 0.87%) and overall liking (−1.15% to 0.82%).

Table 5. Estimates of the carryover effects for different treatments on three attributes (color, taste, and overall liking) using model (2).

Attribute	Parameters [1]	Carryover Effect for Treatment A	Carryover Effect for Treatment B	Carryover Effect for Treatment C
Color	Estimate	+0.10	−0.15	+0.05
	SE	0.08	0.08	0.08
	DF	1074	1074	1074
	t-value	1.31	−1.93	0.62
	Pr > |t|	0.19	0.05	0.53
	% of the scale [2]	+1.29%	−1.90%	+0.61%
Taste	Estimate	+0.03	+0.07	−0.10
	SE	0.09	0.09	0.09
	DF	1074	1074	1074
	t-value	0.34	0.76	−1.09
	Pr > |t|	0.74	0.45	0.27
	% of the scale [2]	+0.39%	+0.87%	−1.26%
Overall Liking	Estimate	+0.03	+0.07	−0.09
	SE	0.09	0.09	0.09
	DF	1074	1074	1074
	t-value	0.30	0.75	−1.05
	Pr > |t|	0.76	0.45	0.29
	% of the scale [2]	+0.33%	+0.82%	−1.15%

[1] SE = Standard error. DF = Degrees of freedom. t-value under the null hypothesis of Ho: Estimate = 0. [2] % of the Scale = Percentage of the estimate on a 9-point Hedonic Scale (Estimate × 100/8).

The carryover effects can be classified as contrast and convergence effects, depending on the valence (sign) of the carryover effects (positive and negative), and the intensity of liking (high, medium, or low liking scores) of the previous and current samples tested by panelists. For example, Table 5 showed a positive carryover effect of treatment A for color as +0.10. In Table 4, the rank of the liking scores for the 3 grape juices was A, B, and C for all sensory attributes. If sample A was followed by sample C in the sequence, the consequence of the positive carryover effect of A would be the inflation of the liking of sample C, and therefore the treatment difference would be smaller than if it had been presented in the first position. In this case, the positive carryover effect (+0.10) of A would be classified as the convergence effect. On the other hand, Table 5 shows a negative carryover effect of sample B for color as −0.15. If sample B was followed by sample C in the sequence, then the consequence of the negative carryover effect of B would be a decrease in the liking of sample C, and therefore the treatment difference would be greater than if it had been presented in the first position. In this case, the negative carryover effect (−0.15) of treatment A would be classified as the contrast effect. Table 6 shows the classification of the carryover effects for the treatments (A, B, and C) and attributes (color, taste, and overall) as evaluated in this study.

Carryover was found to affect the descriptive scores of negative attributes in cheeses [28]. In a descriptive analysis of restructured steaks with 15 sensory attributes, Schlich [30] reported that greasiness, tenderness, and juiciness were affected by the carryover effects of previous samples. Conversely, previous studies found no evidence of carryover effects in beverage products [31,32]. Among the treatment effect for all evaluated attributes (color, taste, and overall liking), the visual attribute (color) may hae been largely responsible for the differences among grape juice samples. Color and visual cues of the samples can affect the expectations of consumer and generate contrast and assimilation (convergence) effects if the product does not match the initial expectations [33]. In previous studies, color has been identified as an attribute that can change taste and flavor perception [34,35]. Color is the first attribute that is evaluated by consumers, and it may have the greatest exposure to carryover effects, since it is the attribute that is initially compared to the previous sample.

The sensory responses in the present study were significantly ($p < 0.05$) and positively correlated ((color-taste, $r = 0.74$), (color-overall, $r = 0.79$), (taste-overall, $r = 0.96$); data not shown). Highly positive correlations among sensory attributes are not uncommon in sensory trials (known as Halo effect). This psychological bias has been also studied [1]. However, acknowledging that the attribute responses may be correlated, the present study showed a useful method to characterize and quantify the carryover effects using the mixed linear model (2).

For model (3), the absolute value of the proportionality parameter estimate (λ) was higher for color than for taste and overall liking (0.16 vs. 0.004 to -0.04) (Table 7). The carryover effect for color was significant (confidence intervals of -0.21 and -0.10). The treatment carryover effect for color was higher (-15.46%) than for taste and overall liking (-0.38 to -3.85%). For all attributes, the proportionality parameter (λ) values were negative; thus, the carryover effect mainly took the form of contrast.

Table 6. Characterization of the carryover effects on three attributes (color, taste, and overall liking) from model (2).

Previous Sample	Present Sample	Carryover Effects		
		Effect on Color	Effect on Taste	Effect on Overall Liking
A	B	Convergence	Convergence	Convergence
A	C	Convergence	Convergence	Convergence
B	A	Convergence	Contrast	Contrast
B	C	Contrast	Convergence	Convergence
C	A	Contrast	Convergence	Convergence
C	B	Contrast	Convergence	Convergence

Models (1)–(3) showed that the treatment carryover effects for color were higher than for taste and overall liking. Only model (3) showed a significant carryover effect (for color). Although there was some disagreement between models (2) and (3) on the form (convergence vs. contrast) of the carryover effects for taste and overall liking, the carryover effects for these attributes were not significant. However, these generated models are useful tools that can be applied to different food and beverage products in sensory trials that use cross-over and Williams experimental designs. Nonetheless, carryover effects cannot be ignored as it may lead to misinterpretation of the results [18]. The effect that a treatment might have on the assessment of the next treatment (carryover) is more likely to occur when using inexperienced consumers than when using trained panelists [18].

Although models (1) and (2) are based on the typical linear mixed models to account for carryover effects, it should be noted that these models treated the ordinal rating variable as continuous and therefore may be subject to a loss of information between the dependent and independent variables. Despite this disadvantage, Long and Freese [36] agreed that ordinal variables are often treated as being continuous for specific experimental designs. Other alternatives such as the contingency table analysis, cumulative logistic regression, and non-parametric methods can be used for evaluating the ordinal dependent variables, although they have limitations such as a strong requirement of proportional odds across all categories, which is often difficult to meet. Unfortunately, these approaches are not readily available in the context of experimental designs. The great advantage of employing linear model approaches in the context of using the appropriate experimental design (cross-over and Williams) may be enough to offset any disadvantages with which they are accompanied. Moreover, the present study may be subject to measurement errors resulting from the participant's judgment bias. Measurement errors in dependent variables do not bias the regression estimate but may increase the standard error of the estimate, which in turn may decrease the power. Fortunately, our large sample size of N = 540 may be helpful to offset this issue.

Table 7. Estimates of the proportionality parameter (λ) for the carryover effects on three attributes (color, taste, overall liking) from model (3).

Attribute	Estimate (λ)	SE [1]	Approximate 95% Confidence Limits		% of Treatment Effect [2]
Color	−0.16	0.03	−0.21	−0.10	−15.46%
Taste	−0.004	0.03	−0.06	0.05	−0.38%
Overall liking	−0.04	0.03	−0.09	0.01	−3.85%

[1] SE = Standard Error. According to Ferris et al. [2], when the estimated λ is positive there is an assimilation of the previous stimulus, while for negative λ carryover takes the form of a contrast effect. The model (N = 540) was fitted using PROC NLIN of the SAS software. [2] % of treatment effect is the percentage of the carryover effect with respect to the treatment effect in the model.

4. Conclusions

Consumer tests are prone to carryover effects where responses of samples being evaluated are affected by previously assessed samples. Although there are some published works regarding the quantification of carryover effects associated with trained assessors, quantifying carryover effects on consumer panelists is still unclear. This study characterized and quantified carryover effects on sensory acceptability of grape juices using mixed and nonlinear models. Results from this study showed that color presented a weak carryover effect among the grape juice samples. Besides, this study also proposed useful modeling tools to characterize and quantify the carryover effects in sensory trials using untrained consumers. Further studies are needed to understand the full extent of carryover effects in different food/beverage products and sensory attributes using consumer panels.

Author Contributions: D.D.T. and W.J. contributed equally to the experiment design, data analysis, and paper writing; J.W., P.C., and S.S. contributed to the experiment layout, data analysis, and revision of the paper; W.P. contributed to the experiment design, interpretation of the results, and paper editing.

Funding: This research was funded partially by the 2017 Early Career Researcher Grant Scheme from the University of Melbourne, Australia (603403) and the USDA NIFA Hatch project LAB94291, Louisiana State University Agricultural Center.

Conflicts of Interest: The authors declare no conflict of interest.

References

1. Lawless, H.T.; Heymann, H. *Sensory Evaluation of Food: Principles and Practices*; Chapman & Hall Press: New York, NY, USA, 1998.
2. Ferris, S.J.; Kempton, R.A.; Muir, D.D. Carryover in sensory trials. *Food Qual. Prefer.* **2003**, *14*, 299–304. [CrossRef]
3. Meilgaard, M.; Civille, G.V.; Carr, B.T. *Sensory Evaluation Techniques*, 4th ed.; CRC Press: Boca Raton, FL, USA, 2007.
4. Ball, R.D. Incomplete block designs for the minimisation of order and carry-over effects in sensory analysis. *Food Qual. Prefer.* **1997**, *8*, 111–118. [CrossRef]
5. Bose, M.; Stufken, J. Optimal crossover designs when carryover effects are proportional to direct effects. *J. Stat. Plan. Inference* **2007**, *137*, 3291–3302. [CrossRef]
6. Clouser-Roche, A.; Johnson, K.; Fast, D.; Tang, D. Beyond pass/fail: A procedure for evaluating the effect of carryover in bioanalytical LC/MS/MS methods. *J. Pharm. Biomed. Anal.* **2008**, *47*, 146–155. [CrossRef] [PubMed]
7. Walter, F.; Boakes, R.A. Long-term range effects in hedonic ratings. *Food Qual. Prefer.* **2009**, *20*, 440–449. [CrossRef]
8. Jirangrat, W.; Wang, J.; Sriwattana, S.; No, H.K.; Prinyawiwatkul, W. The split plot with repeated randomised complete block design can reduce psychological biases in consumer acceptance testing. *Int. J. Food Sci. Technol.* **2014**, *49*, 1106–1111. [CrossRef]
9. Johnson, E.A.; Vickers, Z. The effectiveness of palate cleansing strategies for evaluating the bitterness of caffeine in cream cheese. *Food Qual. Prefer.* **2004**, *15*, 311–316. [CrossRef]
10. Warnock, A.R.; Delwiche, J.F. Regional variation in sweet suppression. *J. Sens. Stud.* **2006**, *21*, 348–361. [CrossRef]

11. Kilcast, D.; Clegg, S. Sensory perception of creaminess and its relationship with food structure. *Food Qual. Prefer.* **2002**, *13*, 609–623. [CrossRef]
12. Lethuaut, L.; Brossard, C.; Rousseau, F.; Bousseau, B.; Genot, C. Sweetness–texture interactions in model dairy desserts: Effect of sucrose concentration and the carrageenan type. *Int. Dairy J.* **2003**, *13*, 631–641. [CrossRef]
13. Rason, J.; Martin, J.F.; Dufour, E.; Lebecque, A. Diversity of the sensory characteristics of traditional dry sausages from the centre of France: Relation with regional manufacturing practice. *Food Qual. Prefer.* **2007**, *18*, 517–530. [CrossRef]
14. Richardson-Harman, N.J.; Stevens, R.; Walker, S.; Gamble, J.; Miller, M.; Wong, M.; McPherson, A. Mapping consumer perceptions of creaminess and liking for liquid dairy products. *Food Qual. Prefer.* **2000**, *11*, 239–246. [CrossRef]
15. Rétiveau, A.; Chambers, D.H.; Esteve, E. Developing a lexicon for the flavor description of French cheeses. *Food Qual. Prefer.* **2005**, *16*, 517–527. [CrossRef]
16. King, E.S.; Dunn, R.L.; Heymann, H. The influence of alcohol on the sensory perception of red wines. *Food Qual. Prefer.* **2013**, *28*, 235–243. [CrossRef]
17. Schifferstein, H.N.; Kuiper, W.E. Sequence effects in hedonic judgments of taste stimuli. *Percept. Psychophys.* **1997**, *59*, 900–912. [CrossRef] [PubMed]
18. MacFie, H.J.; Bratchell, N.; Greenhoff, K.; Vallis, L.V. Designs to balance the effect of order of presentation and first-order carry-over effects in hall tests. *J. Sens. Stud.* **1989**, *4*, 129–148. [CrossRef]
19. Lindstrom, M.J.; Bates, D.M. Nonlinear mixed effects models for repeated measures data. *Biometrics* **1990**, *46*, 673–687. [CrossRef] [PubMed]
20. Davidian, M.; Giltinan, D.M. Nonlinear models for repeated measurement data: An overview and update. *J. Agric. Biol. Environ. Stat.* **2003**, *8*, 387. [CrossRef]
21. Bakker, J.; Brown, W.E.; Hills, B.M.; Boudaud, N.; Wilson, C.E.; Harrison, M. Effect of the food matrix on flavour release and perception. *Spec. Publ. R. Soc. Chem.* **1996**, *197*, 369–374.
22. Cochran, W.G.; Cox, G.M. *Experimental Design*; John Wiley & Sons: New York, NY, USA, 1957.
23. Peryam, D.R.; Pilgrim, F.J. Hedonic scale method of measuring food preferences. *Food Technol.* **1957**, *11*, 9–14.
24. Cordonnier, S.M.; Delwiche, J.F. An alternative method for assessing liking: Positional relative rating versus the 9-point hedonic scale. *J. Sens. Stud.* **2008**, *23*, 284–292. [CrossRef]
25. Lee, H.J.; Kim, K.O.; O'Mahony, M. Effects of forgetting on various protocols for category and line scales of intensity. *J. Sens. Stud.* **2001**, *16*, 327–342. [CrossRef]
26. Laska, E.; Meisner, M.; Kushner, H.B. Optimal crossover designs in the presence of carryover effects. *Biometrics* **1983**, *39*, 1087–1091. [CrossRef] [PubMed]
27. Stone, H. *Sensory Evaluation Practices*; Academic Press: Cambridge, MA, USA, 2012.
28. Muir, D.D.; Hunter, E.A. Sensory evaluation of Cheddar cheese: Order of tasting and carryover effects. *Food Qual. Prefer.* **1991**, *3*, 141–145. [CrossRef]
29. Kofes, J.; Naqvi, S.; Cece, A.; Yeh, M. Understanding presentation order effects and ways to control them in consumer testing. In Proceedings of the 8th Pangborn Sensory Science Symposium, Florence, Italy, 26–30 July 2009.
30. Schlich, P. Uses of change-over designs and repeated measurements in sensory and consumer studies. *Food Qual. Prefer.* **1993**, *4*, 223–235. [CrossRef]
31. Bruvold, W.H. Rated acceptability of mineral taste in water: III. Contrast and position effects in quality scale ratings. *J. Exp. Psychol.* **1970**, *85*, 258. [CrossRef] [PubMed]
32. Mead, R.; Gay, C. Sequential design of sensory trials. *Food Qual. Prefer.* **1995**, *6*, 271–280. [CrossRef]
33. Deliza, R.; MacFie, H.J. The generation of sensory expectation by external cues and its effect on sensory perception and hedonic ratings: A review. *J. Sens. Stud.* **1996**, *11*, 103–128. [CrossRef]

34. Hyman, A. The influence of color on the taste perception of carbonated water preparations. *Bull. Psychon. Soc.* **1983**, *21*, 145–148. [CrossRef]
35. Zampini, M.; Sanabria, D.; Phillips, N.; Spence, C. The multisensory perception of flavor: Assessing the influence of color cues on flavor discrimination responses. *Food Qual. Prefer.* **2007**, *18*, 975–984. [CrossRef]
36. Long, J.S.; Freese, J. *Regression Models for Categorical Dependent Variables Using Stata*, 2nd ed.; Stata Press: College Station, TX, USA, 2006.

© 2018 by the authors. Licensee MDPI, Basel, Switzerland. This article is an open access article distributed under the terms and conditions of the Creative Commons Attribution (CC BY) license (http://creativecommons.org/licenses/by/4.0/).

Communication

What Temperature of Coffee Exceeds the Pain Threshold? Pilot Study of a Sensory Analysis Method as Basis for Cancer Risk Assessment

Julia Dirler [1], Gertrud Winkler [1] and Dirk W. Lachenmeier [2,*]

[1] University of Applied Sciences Albstadt-Sigmaringen, Department Life Sciences, Anton-Günther-Str. 51, 72488 Sigmaringen, Germany; julia.dirler@web.de (J.D.) winkler@hs-albsig.de (G.W.)
[2] Chemisches und Veterinäruntersuchungsamt (CVUA) Karlsruhe, Weissenburger Strasse 3, 76187 Karlsruhe, Germany
* Correspondence: lachenmeier@web.de; Tel.: +49-721-926-5434

Received: 25 April 2018; Accepted: 26 May 2018; Published: 1 June 2018

Abstract: The International Agency for Research on Cancer (IARC) evaluates "very hot (>65 °C) beverages" as probably carcinogenic to humans. However, there is a lack of research regarding what temperatures consumers actually perceive as "very hot" or as "too hot". A method for sensory analysis of such threshold temperatures was developed. The participants were asked to mix a very hot coffee step by step into a cooler coffee. Because of that, the coffee to be tasted was incrementally increased in temperature during the test. The participants took a sip at every addition, until they perceive the beverage as too hot for consumption. The protocol was evaluated in the form of a pilot study using 87 participants. Interestingly, the average pain threshold of the test group (67 °C) and the preferred drinking temperature (63 °C) iterated around the IARC threshold for carcinogenicity. The developed methodology was found as fit for the purpose and may be applied in larger studies.

Keywords: coffee; hot beverages; temperature; esophageal cancer; thermosensing; sensory thresholds; methodological study

---

1. Introduction

Since 2016, the cancer risk in connection to hot beverage consumption has received increased scrutiny from science and consumers alike. The reason for this has been the classification of "very hot beverage consumption" by the International Agency for Research on Cancer (IARC) into group 2A as "probably carcinogenic to humans" [1,2]. Specifically, the risk of developing oesophageal carcinoma increases with the consumption of very hot beverages as shown by a number of epidemiological studies [3–8]. Beverages above 65 °C are considered "very hot" [1,2].

There are only a few studies available that researched the perception of temperature when consuming hot drinks. In general, the thermoreceptors are responsible for the sensation of heat and cold. These receptors are located in the skin and mucous membranes. When an action potential occurs, these receptors relay the stimulus to the nervous system, triggering a sensation [9,10]. The thermoreceptors are located 0.1 to 0.6 mm below the skin surface, in the dermis. These receptors are located not only on the surface of the skin, but also inside the body, e.g., on the internal organs and their mucous membranes [11]. The thermoreceptors can be divided into cold and warm ones. These react during cooling or warming with an impulse increase, thus leading to an action potential. More specifically, transient receptor potential (TRP) channels sense hot and cold. TRP channels respond to stimuli from temperature, pressure, inflammatory agents, and receptor activation. TRP cation channel subfamily V member 1 (TRPV1) receptors open at temperatures greater than 43 °C. TRP cation channel subfamily M member 8 (TRPM8) channels sense cooling, and open at temperatures <26 °C [12,13]. TRPV1 and TRPM8 channels have unusually large Q_{10} values (>15 for TRPV1) [12,14,15].

Important factors in the sensation of temperature are the absolute temperature, the steepness of the change in the temperature which affects the skin during a certain time and the size of the irritated body surface. In addition, the thermal conductivity of the object or the fluid plays a role [11]. Above temperatures of about 44–45 °C, the human begins to develop a painful heat sensation. Pain stimuli are absorbed by pain receptors. These receptors are located in the epithelia of the skin and mucous membranes. These receptors do not approach a specific organ but run in the intercellular clefts of the epithelium. The pain receptors react differently to different stimuli, e.g., on heat, pressure and strain [11]. Only after a series of action potentials and the exceedance of the threshold value, a pain stimulus is triggered [11]. At higher temperatures (surface temperatures of >70 °C for less than one second), trans-epidermal necrosis may occur [16].

The increase in the temperature at the tongue surface results from the contact temperature between the liquid and the tongue. The contact temperature is the point at which two bodies touch each other at different temperatures. This temperature can be estimated by the simple Formula (1) [17]:

$$T_K = T_2 + \frac{b_1}{b_1 + b_2} * (T_1 - T_2) \tag{1}$$

where T_K = Contact temperature, $b_{1/2}$ = Thermal effusivity, $T_{1/2}$ = Temperature of body $\frac{1}{2}$.

The skin temperature is about 37.4 °C with a thermal effusivity of about 1.3 $kWs^{0.5}m^{-2}K^{-1}$. The thermal effusivity of water, on the other hand, is 1.6 $kWs^{0.5}m^{-2}K^{-1}$ [17]. If the coffee has a temperature of 70 °C, thus the contact temperature estimation will be about 55 °C. On the other hand, the formula results in a temperature of 57 °C, which would cause the tongue to be heated at around the pain threshold temperature of 48 °C [18]. This is corroborated by the optimal drinking temperature of 57.8 °C postulated by Brown and Diller [19] based on data modelling for simulating burns from various in vivo studies. These theoretical estimations well confirm the measurements of proband's tongue surfaces of Lee et al. [18]. Recommended maximum temperatures for water to avoid burning were 65 °C for contact periods up to 1 s duration and 60 °C for contact periods of 3–4 s [20].

Only a few experimental approaches are available for the determination of drinking temperatures by means of sensory tests. Graham et al. [21] adapted the method of Pearson and McCloy [22] to estimate the preferred drinking temperature of hot drinks. Hot water, with an initial temperature of 80 to 85 °C, was filled in a porcelain cup. Each time the water cooled down 2 °C, participants were asked to sample the water and give their assessment of the current temperature [21]. Another study also determined which temperatures of the hot drinks are perceived as preferable by consumers. For this purpose, a mixing method of coffee with different temperatures has been developed. The participants were asked to mix their coffee to that temperature they usually would consume the beverage. At first, very hot coffee has been in the cup to be tasted, which was gradually mixed with colder coffee. Having reached the optimum temperature of the coffee, the temperature was measured and documented [23]. Borchgrevink et al. [24] used a similar design but with randomized sequence of temperatures between 57 °C and 91 °C aiming to minimize the potential of an order or undesired treatment effect. In even another design, Pipatsattayanuwong et al. [25] presented six different temperatures ranging from 39 to 82 °C but the two central temperatures (61 °C and 72 °C) were to be tasted first to avoid initial tasting of extremes.

The aim of this study was to develop a method to elucidate which temperatures of hot beverages are perceived as too hot. That is the temperature at which consumers can no longer drink the coffee without feeling pain. Since published reports were open to interpretation, we developed a new method based on the study of Lee and O'Mahony [23] but with an inverse experimental design because we judged it as being inappropriate to start from the pain stimulus directly in the very first tasting. The randomized design of Borchgrevink et al. [24] would also not completely avoid this effect as some participants may still begin with potentially scalding beverages with the potential to make them unable to adequately assess the following lower temperature beverages. Therefore, in our design the temperatures of coffee are gradually increased until the pain stimulus level is reached. For a pilot study of the method, coffee (standard caffeine-containing type) is used as model because it is the most commonly consumed hot beverage in Germany and no data at all regarding this question is available for the German population.

2. Materials and Methods

The mixing method according to Lee and O'Mahony [23] is used with some modifications. In contrast to the reduction of temperature in the original protocol, the coffee temperature is gradually increased by adding very hot coffee until the participants perceive the drink as being too hot. In each step, the participants re-test the hot drink after a temperature increase of 2–3 °C.

The experimental setup is shown in Figure 1. Each participant receives a cup of cold water, a cup as spittoon, a thermometer (Testo 108, Testo, Lenzkirch, Germany), a beaker, a thermos flask with very hot coffee (Instant coffee, Brand "Gut und Günstig", Edeka, Hamburg, Germany) and an isolation cup (Styrofoam cup, 400 mL Thermo Cup EPS Neutral, white, Gastro-Sun.de, Blankenhain, Germany) with the "colder" coffee.

Before the start of the study, the participants are trained through a short briefing so that all of them have understood the test procedure and any questions that may arise can be clarified before the start of the test. The test sheet (Figures A1 and A2, Appendix A), which is given to the participant to be filled in during the test run, also contains the test instructions. Then it contains a table, in which the temperatures of the tasted coffees are entered. For each tasted temperature of the hot drink, the participants give a judgement about the taste sensation using a three-category scale (with the word descriptors: too cold, optimal, or too hot) and may make a brief remark in their own words. Finally, the participants indicate how often they usually ingest hot drinks.

The coffee is prepared each time with the same amount of tap water (200 mL) and soluble coffee powder (4.5 g). To ensure a similar starting temperature for all participants, a pot of coffee is prepared 15 min before they arrive. This coffee is cooled down to the desired temperature using a thermostatted water bath with temperature control. The desired starting temperature for all participants is 60 °C.

The very hot coffee, however, is poured only shortly before the start of the experiment to ensure its high temperature. In order to be able to hinder the rapid cooling of the coffee, tests are carried out with a cup holder made of foam (Florence foam holder for 12 glasses/cups, 330 × 245 × 60 mm, Haba BV, Maasland, The Netherlands). With the help of the foam, the cups are isolated and thus delay the cooling. In order to minimize the cooling of the coffee in the thermos flasks as well, they are preheated before the experiments by adding boiling hot water. The water is removed from the thermos flasks shortly before the test persons arrive. By using a water dispenser (Bunn H3EA Hot Water Dispenser, Bunn, Springfield, IL, USA) for the preparation of the "very hot" coffee, constant temperatures of about 96 °C can be achieved. This coffee is prepared only after the brief introduction of the participants to keep these temperatures as long as possible.

The participants gradually mix in the test phase in an ascending fashion very hot coffee (about 96 °C) to the colder coffee (starting temperature 60 °C, 150 mL) and measure the temperature of the mixture. The added amount is standardized by a small beaker (0.04 L), which leads to a temperature increase of 2–3 °C. Then the coffee is sampled. The process of mixing is repeated by the participants until their personal optimum drinking temperature is reached, followed by the temperature, which is perceived as unpleasant, painful or too hot at which the trial is stopped. The duration of the trial is about 15–25 min.

For a pilot feasibility test of the study protocol, all students, staff and professors of the Albstadt-Sigmaringen University were invited by e-mail to participate. A total of 87 Caucasian people (65 female and 22 male; average age 33 ± 15) and thus about 5% of the university location Sigmaringen members participated in the study. The informed consent of the participants was obtained with signatures in writing following an oral information about the trial. The study protocol was approved by the ethics commission at Landesärztekammer Baden-Württemberg (Stuttgart, Germany) at 19 December 2017, Az. F-2017-094. Data was tested for normality using Shapiro–Wilk test and group comparisons were conducted using one-way analysis of variance (Minitab 16, Minitab Inc., State College, PA, USA). Statistical significance was assumed at the 0.05 significance level.

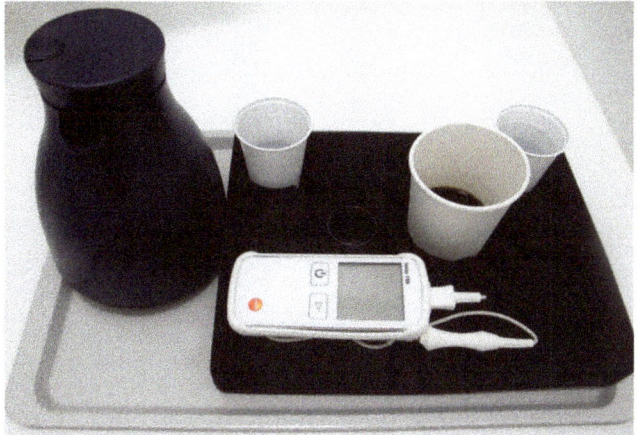

Figure 1. Experimental set-up provided to each participant. The isolation cup contains 150 mL of coffee at 60 °C and the thermos flask coffee at 93 °C. The small 40 mL beaker cup is used to gradually add the hot coffee. Two further cups with cold water for spilling if required and an empty cup for spitting are provided. The digital thermometer was used to measure the temperature in each tasting step.

3. Results

The raw results of each participant are presented in Appendix B, Table A1. Descriptive statistical analysis shows that on average between participants, coffee is perceived as "too hot" from temperatures beyond 66 °C (Table 1). The standard deviation is 3 °C, which appears to be comparably low in the context of sensory analysis. The highest mentioned temperature for the pain threshold of one of the participants is 71 °C and is thus well above the specified IARC threshold of 65 °C. The range shows that the participants have a very different perception of the threshold temperature. Temperatures from 58 °C to 71 °C are determined as maximum tolerable temperatures. This results in a span of 13 °C for the pain threshold.

Table 1. Summarized results of the pilot study (n = 87).

Temperature	Mean	Standard Deviation	Minimum	Maximum	Range	Median
Pain threshold [a]	66 °C	3 °C	58 °C	71 °C	13 °C	66 °C
Preferred drinking temperature	63 °C	3 °C	55 °C	70 °C	15 °C	63 °C

[a] Temperature perceived as "too hot".

Figure 2 shows the distribution of the temperatures perceived as "too hot". Again, it can be seen that most of the participants' responses for the pain threshold are at temperatures around 66 °C to 68 °C (n = 31; 36%). The category of pain threshold temperatures above 68 °C includes 19 participants (22%). Of the 87 participants, 24 persons (28%) already consider temperatures below 65 °C to be "too hot"; one participant already considered a temperature below 60 °C as being "too hot" (1%).

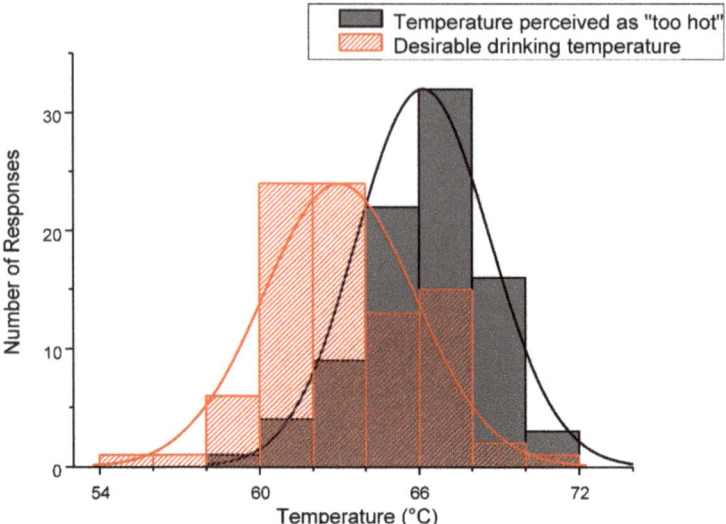

Figure 2. Histogram of the distributions for temperatures perceived as "too hot" and as "desirable/preferred" ($n = 87$). The curves show the normal distribution for both data sets.

In addition to the maximum temperature perceived as "too hot", the study also determined which temperatures the participants feel most desirable or preferable for coffee consumption. The descriptive statistical analysis gives a mean value of 63 °C with a standard deviation of 3 °C. The values range from a minimum of 55 °C to a maximum of 70 °C. Among the respondents are 19 participants (22%) who find their coffee at temperatures above 65 °C to be optimally temperated. Among them are two participants (2%) who even find the drinking desirable at temperatures above 68 °C. However, of all participants, a large majority of 63 persons (72%) has their personal drinking temperature below 65 °C.

The distributions for both temperatures (Figure 2) are normal. The population means are significantly different. The average difference between the desirable and pain threshold temperatures was 3 °C (standard deviation 2 °C).

4. Discussion

During an initial trial, the rapid cooling of the coffee and thus a difficulty of gradually increasing the temperature of the beverage was observed, especially when normal porcelain coffee cups were used. Based on previous research on cooling of coffee in different materials [26,27], the experimental setup was modified to include insulation in a foam cup holder and the use of a styrofoam cup. The cups remain in the foam holder during the experiment, only for drinking the cup is removed. Furthermore, it is important that the hot coffee for increasing the temperature is made as hot as possible, in our case using preheated thermos flasks and a hot water dispenser. Through these measures, it has been possible to almost linearly increase the temperature in the cup. Previous research has shown that the cooling behaviour of coffee is identical to the one of hot water [27], so that the study design may be transferrable to most hot beverages with similar behaviour. However, there could be an influence or interaction between acidic beverages (such as coffee) and temperature sensation, because the TRPV1 receptor integrates stimuli including heat and extracellular acidic pH [28].

Our study protocol was well usable for its purpose and no inconsistencies were observed besides one participant who already judged the initial coffee as "too hot", but this problem can be circumvented by allowing the coffee to cool for some minutes or by decreasing the initial temperature.

According to the literature, there may be a considerable intra-individual variation in pain threshold tests (12 °C on average [23]). While such differences were not researched in our study (only one trial per participant), the inter-individual variation in our trial is lower than this intra-individual value from the literature. This can potentially be explained by differences in the experimental design (starting from hot to cold in [23], or presenting only very few samples with large temperature differences in Refs. [24,25]). We believe that large receptor responses such as pain may hinder or restrict taste testing for some time, so that the starting point should be the low stimulus and not the high stimulus (similar to, e.g., standard procedures for taste tests for salt, acidity, etc., see ISO 8586 [29], which always start at the lowest concentration). Final evidence into this question may be only gained by testing the same beverages with the same collectives in both fashions (i.e., increasing and decreasing temperatures). This could also help to investigate the question if a hysteresis effect may occur due to the increasing temperature in the experimental design.

The strength of coffee may also play a role in the sensation of temperature [23]. However, only one single coffee type was used in this study and the coffee in the study by Lee and O'Mahony [23] was much stronger. Coffee type may therefore potentially explain the larger variances in Lee and O'Mahony [23] compared to this study. A larger bitterness response (or pH decrease, see above), for example, may interact with the temperature response. Interactions may occur between irritant stimuli (heat or cold) and the human taste system either at the peripheral level (receptor level which is unlikely), or the central nervous system level.

In addition, this study does not assess whether there are differences between consumers drinking their coffee with milk/cream/sugar or black without additions. Therefore, the protocol could be expanded in the future by including tastants (such as sourness and bitterness in the coffee) and flavourants. While the perceived intensity of basic tastes does not strictly increase with temperature, this is true for some flavours. Caffeine is also a bioactive compound, which may lead to individual reactions due to genetics (the influence of caffeine could be tested by comparison with decaffeinated coffee). Before further use, the protocol should be validated as to whether reproducible results are achieved both intra-individual and inter-individual within taste-testing panels and for different beverage types and preparations. A potential limitation of the protocol may also be the three-category scale for descriptors, which could possibly be expanded using a larger number of categories or a numeric pain rating scale.

The results of the pilot test study show that, on average, the participants judge coffee as being "too hot" just above the threshold of 65 °C suggested by IARC for "very hot" beverages that are probably carcinogenic to humans [1,2]. This consumption preference of German consumers may explain the fact that an increased incidence of oesophageal cancer has not been described in connection to hot beverage consumption in Germany, while in other countries, such as for tea in Iran or mate in South America, where such epidemiological associations were mainly described, the preferred consumption temperature was typically much higher than 70 °C [3–8]. All other previous studies were conducted in the USA, and very similar consumer preference was detected in this country as in our study (ideal consumption temperature 63–68 °C [24]; optimal drinking temperature 58 °C [19]; most preferred serving temperature 75 °C [25]; mean preferred temperature for drinking 60 °C [23]; preferred temperature of ingestion 56 °C [22]).

Nevertheless, some participants in Germany tolerate temperatures well above 70 °C and thus may be exposed to an increased risk of oesophageal cancer. Therefore, the question arises how this group of consumers can tolerate such high temperatures without experiencing pain.

According to Lee et al. [18], the threshold for sensation of pain on the tongue starts at temperatures of about 46–48 °C, while the TRPV1 channels open at 43 °C. However, the average coffee drinking temperatures in our study clearly exceed this level. Obviously, the liquid has much higher temperatures than the surface of the tongue when heated by the coffee. But measurements of Lee et al. [18] showed that coffee at a temperature of 60 °C may raise the surface of the tongue to 53 °C, which is above the postulated pain threshold of the tongue.

Still, many of the participants in our study perceive coffee temperatures of around 60 °C and also much higher as desirable and certainly not too hot. This discrepancy is explained by Lee et al. [18] by the fact that the temperatures may not be kept in the mouth long enough. A swallowing process lasts only a few seconds, so the temperature exposure acting on the mucous membranes in the mouth and on the tongue is not sufficiently long and thus not able to reach an action potential. Thus, no pain is transmitted despite high temperatures, so that consumers are able to drink the hot drinks without pain.

Finally, another hypothesis is that the pain potential can possibly be reduced by a habituation effect through continuous hot beverage consumption over the lifetime [30]. Lütgendorff-Gyllenstorm [31] assumes that the defence mechanism or instinct against consumption of very hot foods and beverages, still present in infancy, is lost during growing up to adulthood by education (as being the desirable consumption behaviour) and gradual adaptation to increased temperatures. Hence, this habituation effect may also explain the large inter-individual differences of up to 15 °C.

Some previous studies considered the question whether temperatures of very hot beverages actually influence cancer risk or other disease outcomes [1,3,21,22]. In fact, previous experimental studies have shown that some consumers may ingest their coffee at temperatures that can burn the epithelium of the oesophagus [23]. It has been shown that the participants on average perceive coffee most preferable at temperatures of about 60 °C. People who drink their coffee black prefer slightly higher temperatures than participants who drink their coffee with milk. Similarly, "weaker" (less concentrated) coffee is preferred at higher temperatures [23]. Our findings basically corroborate these observations, suggesting that the temperature at which coffee is considered preferable in Germany is typically <65 °C, and hence below the threshold for carcinogenic risk [1,2].

5. Conclusions

The developed method is well-suitable for the purpose to obtain temperature preferences of beverages and next steps could include intra- and inter-individual validation work and field testing with larger collectives of participants. After final validation, the method could be useful for testing temperature threshold differences in oesophageal cancer patients versus a control group.

Author Contributions: J.D., G.W. and D.W.L. conceived and designed the experiments; J.D. performed the experiments; J.D. analysed the data; J.D. wrote the initial draft in the form of a Bachelor thesis; D.W.L. summarized the draft as a scientific paper and translated into English; G.W. revised the final draft; J.D., G.W. and D.W.L. revised and approved the final version.

Conflicts of Interest: The authors declare no conflict of interest.

Appendix A

Ab welcher Trinktemperatur werden Heißgetränke als zu heiß empfunden? Untersuchung zum Schmerzschwellenwert als Grundlange für eine Krebsrisikoanalyse.

- Versuchsbogen -

Mein Geschlecht: weiblich: O männlich: O   Mein Geburtsjahr: _____

Anleitung zur Durchführung des aufsteigenden Mischverfahrens: Sie selbst erhöhen bitte stufenweise die Trinktemperatur des Kaffees in Ihrem Trinkbecher um jeweils ca. 2 - 3 °C.

- Dazu geben Sie jeweils ein kleines Becherglas voll heißem Kaffee zum kälteren Kaffee in Ihrem Trinkbecher.
- Lesen Sie die neue Temperatur am Thermometer ab und notieren Sie sie in der Tabelle unten.
- Probieren Sie nun vorsichtig einen kleinen Schluck des Kaffees im Trinkbecher. Sie müssen den Kaffee nicht schlucken, sondern können ihn nach dem Probieren ausspucken. Notieren Sie nun durch Ankreuzen in der Tabelle Ihre Empfindung.

Temperatur in °C						
Wahrnehmung	o zu kalt o optimal o zu heiß	o zu kalt o optimal o zu heiß	o zu kalt o optimal o zu heiß	o zu kalt o optimal o zu heiß	o zu kalt o optimal o zu heiß	o zu kalt o optimal o zu heiß
Kommentar, Notiz						

Temperatur in °C						
Wahrnehmung	o zu kalt o optimal o zu heiß	o zu kalt o optimal o zu heiß	o zu kalt o optimal o zu heiß	o zu kalt o optimal o zu heiß	o zu kalt o optimal o zu heiß	o zu kalt o optimal o zu heiß
Kommentar, Notiz						

Meine optimale Trinktemperatur liegt bei _____ °C

So häufig trinke ich üblicherweise Heißgetränke:
- o > 4 Tassen pro Tag
- o 2-3 Tassen pro Tag
- o 1 Tasse pro Tag
- o seltener

Figure A1. Experimental sheet for each participant of the trial (original in German).

What temperature of hot beverages is perceived as too hot? Pain threshold analysis as basis for cancer risk assessment.

My sex: female: O male: O   Year of birth: _____

Instructions for carrying out the ascending mixing procedure: You yourself will gradually increase the drinking temperature of the coffee in your cup by approximately 2 - 3 °C.
- Add a small beaker full of hot coffee to the colder coffee in your cup.
- Read the new temperature on the thermometer and record it in the table below.
- Carefully try a small sip of the coffee in the cup. You do not have to swallow the coffee but can spit it out after testing. Finally note your sensation by ticking the appropriate field in the table.

Temperature in °C						
Sensation	o too cold o optimal o too hot	o too cold o optimal o too hot	o too cold o optimal o too hot	o too cold o optimal o too hot	o too cold o optimal o too hot	o too cold o optimal o too hot
Comment, note						

Temperature in °C						
Sensation	o too cold o optimal o too hot	o too cold o optimal o too hot	o too cold o optimal o too hot	o too cold o optimal o too hot	o too cold o optimal o too hot	o too cold o optimal o too hot
Comment, note						

My optimal drinking temperature is _____ °C

My frequency of hot beverage consumption:
- o > 4 cups per day
- o 2-3 cups per day
- o 1 cup per day
- o seldom (<1 cup per day)

Figure A2. Experimental sheet for each participant of the trial (English translation).

Appendix B

Table A1. Raw results of pilot study ($n = 87$).

Gender	Age	Temperature Perceived as "Too Hot" (°C)	Preferred Drinking Temperature (°C)	Coffee Consumption Behaviour (Cups per Day)
male	55	58.0	55.0	2–3
female	45	60.2	<60.0	2–3
male	56	60.2	57.4	>4
female	28	61.5	59.7	2–3
male	35	61.9	58.4	>4
male	24	62.0	60.0	2–3
male	49	62.0	60.0	2–3
female	22	62.4	59.0	1
female	21	63.0	59.6	2–3
male	28	63.0	60.5	2–3
female	22	63.2	61.2	>4
female	19	63.3	61.7	seldom
female	27	63.5	60.8	2–3
female	23	63.8	61.0	2–3
female	34	64.0	61.0	2–3
female	21	64.1	62.2	2–3
female	39	64.1	60.4	2–3
female	23	64.2	62.4	1
male	25	64.2	62.0	2–3
male	63	64.3	62.3	2–3
female	21	64.5	61.7	seldom
male	55	64.5	60.0	>4
female	34	64.8	62.6	2–3
female	38	64.9	58.0	>4
female	24	65.0	63.0	2–3
male	21	65.0	62.5	2–3
female	20	65.2	63.6	seldom
male	72	65.2	61.7	>4
male	22	65.3	63.3	2–3
female	21	65.4	62.5	2–3
female	22	65.4	61.9	>4
female	22	65.6	61.0	2–3
female	23	65.7	62.0	2–3
female	52	65.7	60.2	>4
female	26	65.8	63.0	>4
female	53	65.9	59.3	seldom
female	59	66.0	60.0	>4
male	22	66.0	61.5	2–3
female	26	66.1	60.0	seldom
female	36	66.2	64.0	2–3
female	22	66.3	63.5	2–3
male	53	66.3	60.0	>4
female	22	66.4	61.0	>4
female	23	66.4	64.5	1
female	24	66.5	63.0	2–3
female	41	66.7	60.5	2–3
female	25	66.9	63.0	>4
female	23	67.0	62.0	1
female	19	67.1	65.0	2–3
female	20	67.1	63.4	2–3
female	51	67.1	64.0	2–3
male	52	67.1	61.0	>4
female	21	67.3	65.0	>4
female	23	67.3	66.0	2–3
female	25	67.3	64.5	>4
female	51	67.3	66.0	>4
male	35	67.3	65.0	>4
male	42	67.3	65.0	2–3
female	25	67.4	66.0	1
male	42	67.4	60.0	>4

Table A1. *Cont.*

Gender	Age	Temperature Perceived as "Too Hot" (°C)	Preferred Drinking Temperature (°C)	Coffee Consumption Behaviour (Cups per Day)
female	24	67.5	67.3	>4
female	19	67.6	65.0	seldom
female	24	67.6	66.0	2–3
female	48	67.7	63.3	2–3
female	55	67.8	63.7	>4
male	22	67.8	63.4	2–3
female	54	67.9	67.4	>4
female	62	67.9	63.8	2–3
female	58	68.0	63.3	2–3
female	22	68.2	64.6	2–3
female	21	68.3	64.5	2–3
male	64	68.4	62.8	>4
male	20	68.5	66.0	2–3
female	51	68.6	67.5	2–3
female	22	68.8	68.2	>4
female	56	68.9	67.0	2–3
female	21	69.0	66.0	2–3
female	24	69.0	63.2	2–3
female	24	69.0	64.0	seldom
female	48	69.3	67.0	>4
female	19	69.5	66.1	2–3
female	23	69.5	66.0	2–3
female	29	69.6	67.7	>4
female	22	69.8	67.6	2–3
female	55	70.2	69.3	2–3
male	30	70.5	65.6	>4
female	23	71.2	70.0	>4

Data sorted according to temperature perceived as "too hot".

References

1. IARC Working Group on the Evaluation of Carcinogenic Risks to Humans. Coffee, mate, and very hot beverages. *IARC Monogr. Eval. Carcinog. Risks Hum.* **2018**, *116*. in press.
2. Loomis, D.; Guyton, K.Z.; Grosse, Y.; Lauby-Secretan, B.; El, G.F.; Bouvard, V.; Benbrahim-Tallaa, L.; Guha, N.; Mattock, H.; Straif, K.; et al. Carcinogenicity of drinking coffee, mate, and very hot beverages. *Lancet Oncol.* **2016**, *17*, 877–878. [CrossRef]
3. Islami, F.; Boffetta, P.; Ren, J.S.; Pedoeim, L.; Khatib, D.; Kamangar, F. High-temperature beverages and foods and esophageal cancer risk—A systematic review. *Int. J. Cancer* **2009**, *125*, 491–524. [CrossRef] [PubMed]
4. Ghadirian, P. Thermal irritation and esophageal cancer in northern Iran. *Cancer* **1987**, *60*, 1909–1914. [CrossRef]
5. Andrici, J.; Eslick, G.D. Hot Food and Beverage Consumption and the Risk of Esophageal Cancer: A Meta-Analysis. *Am. J. Prev. Med.* **2015**, *49*, 952–960. [CrossRef] [PubMed]
6. Munishi, M.O.; Hanisch, R.; Mapunda, O.; Ndyetabura, T.; Ndaro, A.; Schuz, J.; Kibiki, G.; McCormack, V. Africa's oesophageal cancer corridor: Do hot beverages contribute? *Cancer Causes Control* **2015**, *26*, 1477–1486. [CrossRef] [PubMed]
7. Okaru, A.O.; Rullmann, A.; Farah, A.; Gonzalez de Mejia, E.; Stern, M.C.; Lachenmeier, D.W. Comparative oesophageal cancer risk assessment of hot beverage consumption (coffee, mate and tea): The margin of exposure of PAH vs. very hot temperatures. *BMC Cancer* **2018**, *18*, 236. [CrossRef] [PubMed]
8. Yu, C.; Tang, H.; Guo, Y.; Bian, Z.; Yang, L.; Chen, Y.; Tang, A.; Zhou, X.; Yang, X.; Chen, J.; et al. Effect of hot tea consumption and its interactions with alcohol and tobacco use on the risk for esophageal cancer: A population-based cohort study. *Ann. Intern. Med.* **2018**, *168*, 489–497. [CrossRef] [PubMed]
9. Filingeri, D. Neurophysiology of Skin Thermal Sensations. *Compr. Physiol.* **2016**, *6*, 1429. [CrossRef] [PubMed]

10. Schmelz, M. Neuronal sensitivity of the skin. *Eur. J. Dermatol.* **2011**, *21* (Suppl. 2), 43–47. [CrossRef] [PubMed]
11. Mörike, K.D.; Betz, E.; Mergenthaler, W. *Biologie des Menschen*; Quelle & Meyer: Wiesbaden, Germany, 2001.
12. Tominaga, M. The Role of TRP Channels in Thermosensation. In *TRP Ion Channel Function in Sensory Transduction and Cellular Signaling Cascades*; Liedtke, W.B., Heller, S., Eds.; CRC Press: Boca Raton, FL, USA, 2007.
13. Venkatachalam, K.; Montell, C. TRP channels. *Annu. Rev. Biochem.* **2007**, *76*, 387–417. [CrossRef] [PubMed]
14. Raddatz, N.; Castillo, J.P.; Gonzalez, C.; Alvarez, O.; Latorre, R. Temperature and voltage coupling to channel opening in transient receptor potential melastatin 8 (TRPM8). *J. Biol. Chem.* **2014**, *289*, 35438–35454. [CrossRef] [PubMed]
15. Ito, E.; Ikemoto, Y.; Yoshioka, T. Thermodynamic implications of high Q 10 of thermo-TRP channels in living cells. *Biophysics (Nagoya-shi)* **2015**, *11*, 33–38. [CrossRef] [PubMed]
16. Moritz, A.R.; Henriques, F.C. Studies of Thermal Injury: II. The Relative Importance of Time and Surface Temperature in the Causation of Cutaneous Burns. *Am. J. Pathol.* **1947**, *23*, 695–720. [PubMed]
17. Baehr, H.D.; Stephan, K. *Wärme- und Stoffübertragung*; Springer: Berlin/Heidelberg, Germany, 2013.
18. Lee, H.S.; Carstens, E.; O'Mahony, M. Drinking hot coffee: Why doesn't it burn the mouth? *J. Sens. Stud.* **2003**, *18*, 19–32. [CrossRef]
19. Brown, F.; Diller, K.R. Calculating the optimum temperature for serving hot beverages. *Burns* **2008**, *34*, 648–654. [CrossRef] [PubMed]
20. Siekmann, H. Recommended maximum temperatures for touchable surfaces. *Appl. Ergon.* **1990**, *21*, 69–73. [CrossRef]
21. Graham, D.Y.; Abou-Sleiman, J.; El-Zimaity, H.M.; Badr, A.; Graham, D.P.; Malaty, H.M. Helicobacter pylori infection, gastritis, and the temperature of choice for hot drinks. *Helicobacter* **1996**, *1*, 172–174. [CrossRef] [PubMed]
22. Pearson, R.C.; McCloy, R.F. Preference for hot drinks is associated with peptic disease. *Gut* **1989**, *30*, 1201–1205. [CrossRef] [PubMed]
23. Lee, H.S.; O'Mahony, M. At what temperatures do consumers like to drink coffee? Mixing methods. *J. Food Sci.* **2002**, *67*, 2774–2777. [CrossRef]
24. Borchgrevink, C.P.; Susskind, A.M.; Tarras, J.M. Consumer preferred hot beverage temperatures. *Food Qual. Prefer.* **1999**, *10*, 117–121. [CrossRef]
25. Pipatsattayanuwong, S.; Lee, H.S.; Lau, S.; O'Mahony, M. Hedonic R-index measurement of temperature preferences for drinking black coffee. *J. Sens. Stud.* **2007**, *16*, 517–536. [CrossRef]
26. Verst, L.-M.; Winkler, G.; Lachenmeier, D.W. Dispensing and serving temperatures of coffee-based hot beverages. Exploratory survey as a basis for cancer risk assessment. *Ernahrungs Umschau* **2018**, *65*, 64–70. [CrossRef]
27. Langer, T.; Winkler, G.; Lachenmeier, D.W. Untersuchungen zum Abkühlverhalten von Heißgetränken vor dem Hintergrund des temperaturbedingten Krebsrisikos. *Deut. Lebensm. Rundsch.* **2018**, *114*. in press.
28. Dhaka, A.; Uzzell, V.; Dubin, A.E.; Mathur, J.; Petrus, M.; Bandell, M.; Patapoutian, A. TRPV1 is activated by both acidic and basic pH. *J. Neurosci.* **2009**, *29*, 153–158. [CrossRef] [PubMed]
29. ISO. *ISO 8586:2012. Sensory Analysis—General Guidelines for the Selection, Training and Monitoring of Selected Assessors and Expert Sensory Assessors*; International Organization for Standardization: Geneva, Switzerland, 2012.
30. Riedl, R.; Lütgendorff-Gyllenstorm, H. Zu heißes Essen und Trinken—Die unbeachtete Gefahr. *Deut. Lebensm. Rundsch.* **2011**, *107*, 71B–83B.
31. Lütgendorff-Gyllenstorm, H. *Risikofaktor Nahrung mit mehr als 37 Grad Celsius. Wir essen und trinken zu heiß*; Verlag Wilhelm Maudrich: Vienna, Austria, 1994.

 © 2018 by the authors. Licensee MDPI, Basel, Switzerland. This article is an open access article distributed under the terms and conditions of the Creative Commons Attribution (CC BY) license (http://creativecommons.org/licenses/by/4.0/).

Review

Application of Sensory Descriptive Analysis and Consumer Studies to Investigate Traditional and Authentic Foods: A Review

Jiyun Yang and Jeehyun Lee *

Department of Food Science and Nutrition, Pusan National University, Busan 26241, Korea; jiyunyang@pusan.ac.kr
* Correspondence: jeehyunlee@pusan.ac.kr; Tel.: +82-51-510-2784

Received: 7 January 2019; Accepted: 14 January 2019; Published: 2 February 2019

Abstract: As globalization progresses, consumers are readily exposed to many foods from various cultures. The need for studying specialty and unique food products, sometimes known as traditional, authentic, ethnic, exotic, or artisanal foods, is increasing to accommodate consumers' growing demands. However, the number of studies conducted on these types of products with good quality sensory testing is limited. In this review, we analyzed and reviewed sensory and consumer research on specialty and unique food products. Various factors such as manufacturing, processing, or preparation methods of the samples influence the characteristics of food products and their acceptability. Sensory descriptive analysis can be used to distinguish characteristics that highlight these differences, and consumer research is used to identify factors that affect acceptability. Familiarity with product attributes contributes to consumer acceptance. When cross-cultural consumer research is conducted to support product market placement and expansion, sensory descriptive analysis should be conducted in parallel to define product characteristics. This allows better prediction of descriptors that influence consumer acceptability, leading to appropriate product modification and successful introduction.

Keywords: sensory evaluation; specialty food; unique food products; ethnic food; descriptive analysis; consumer test

1. Introduction

In the field of food science, sensory science constitutes a discipline dealing with human sensory perceptions and affective responses to various kinds of foods, beverages, and their components that evolved from the need for scientifically sound and systematic sensory evaluation [1]. The conception of sensory science has been attributed to the development of consumer or hedonic food acceptance methodologies that were established in the 1940s by the U.S. Army Corps of Engineers [2]. More recently, sensory science has been defined as "a scientific method used to evoke, measure, analyze, and interpret those responses to products as perceived through the senses of sight, smell, touch, taste and hearing" [3]. Depending on the subject of the sensory science research, various methods may be used; among the sensory evaluation methods, sensory descriptive analysis and consumer acceptability testing are the most frequently used.

Sensory descriptive analysis involves the discrimination and description of both qualitative and quantitative sensory factors of products by trained panels [4,5]. For example, the Flavor Profile [6], Texture Profile [7], Spectrum^{TM} method [4], and Quantitative Descriptive Analysis (QDA$^®$) [8] can be applied as sensory descriptive analysis methods. By using these kinds of methods, it is possible to pinpoint differences among product variants, conditions, identify drivers of consumer hedonic responses, and examine relationships between sensory and chemical characteristics [1,9].

Numerous studies related to the sensory descriptive analysis of food products or sensory methodologies have been published and the use of sensory methods related to product research and development has been described [10]. However, the number of papers reporting the results of good quality sensory descriptive analysis of ethnic, specialty, and exotic food products reflecting traditional and authentic food cultures is limited. The concept of "traditional" food is that of foods that represent a group of people, knowledge, and even local resources [11]. This uniqueness often is referred to as "authenticity", which may be more often used for cultural products, rather than those that are part of our daily routines [12]. Food from different regions or countries can provide "unique" and "exotic" characteristics to individuals from other cultural backgrounds [13]. This often is regarded as "ethnic" foods outside the place of origin and when it is unfamiliar to specific individuals [11,14]. Because of the various possible definitions, in this review we have used the words "traditional", "authentic", "ethnic", "exotic", "unique", and "specialty" interchangeably.

Ethnic food can be defined narrowly as foods originating from a heritage and culture of an ethnic group who use their knowledge of local ingredients. More broadly, ethnic or traditional foods are representative of a cuisine of an ethnic group or country that is culturally and socially distinct and whose foods may be accepted by consumers outside of the respective ethnic group [15]. Ethnic foods provide consumers from other culinary traditions with opportunities to experience new cultures and cuisines [16]. For example, green tea [17,18], rooibos tea [19], fermented soybean [20,21], kimchi [22,23], Portuguese cooked blood sausage [24], and turrón [25] are kinds of ethnic foods that have been analyzed by researchers using sensory descriptive analysis for understanding their attributes.

Consumer acceptability represents one of the most important tests for sensory analysis and often involves a scaling method to measure the degree of liking or disliking of products using naive consumers [3]. However, the degree of acceptability does not constitute the only aspect of consumer studies [26]; consumer emotions, perception, and the relationship between consumers' feelings about a product and descriptive sensory characteristics and instrumental information can also be determined [27]. Among those, food acceptance constitutes an essential outcome of the interaction between humans and foods [28]. Food acceptance may be affected by food habits, attitude, and beliefs [29,30], with culture (i.e., tradition) serving as one primary factor that underlies food choices [31,32]. Differences in the food environment and dietary experience across cultures may influence the preference for sensory characteristics of food products [33]. Similarly, familiarity with food products also may affect food choice [34] and food beliefs and potential acceptability [35]. Numerous consumer studies have been conducted on the acceptability of commercial food products as well as categories of food and various consumer methods, perceptions, emotions, and cross-cultural studies [36–40]. Although less prevalent, some consumer studies have also investigated traditional or unique food products such as traditional cheese [41–45], Doenjang (Korean traditional fermented soybean paste) [46,47], açai juice [48], polenta sticks [49], Argentinean fermented sausages [50], olive oil [51,52], and Korean traditional soup [53].

As globalization progresses, it has become easier to participate in and appreciate other cultures and many cultural traditions are being shared. This has led to an increasing number of food-savvy consumers and the trend for once-unfamiliar cuisines and flavors to become basic and standard elements in day-to-day diets of other cultures [54]. In particular, younger consumers are already accustomed to certain food products that may have been regarded previously as unique or novel. For example, concomitant with the increased exposure to ethnic or traditional foods, consumers enthusiastically use various kinds of spices to enhance the flavor of their dishes. In particular, in the U.S. market, spice consumption has grown almost three times as fast as the population over the last few decades [54,55]. The importance of not only spices however also other unique or special food products is growing in accordance with globalization and consumer needs. However, many more such foods exist than those that have already attained general awareness.

It is necessary to characterize these unique international food products to better understand and accommodate consumer demands for unfamiliar food products. In particular, no reviews have been published on sensory studies regarding such food products. The aim of the present review is, therefore, to describe and summarize the high-quality descriptive studies and consumer research that is currently available for special and unique food products such as traditional or ethnic foods. Accordingly, we will compare the conducted research that is based on the following perspectives: sample type and number and methods or procedure of descriptive and consumer studies. Data analysis and results will also be briefly compared between studies.

2. Literature Review

Articles were searched from the web-library of Pusan National University (https://lib.pusan.ac.kr) using keywords. In the case of sensory descriptive analysis, "descriptive sensory lexicon", "descriptive sensory terminology", and "descriptive sensory characteristics" were searched from 2000 to 15 August 2018. Consumer studies were searched using "consumer liking" for the same time period. Hundreds of papers were searched because of the wide range of keywords, although only a limited number of these evaluated traditional or ethnic foods as samples. All selected studies comprised research papers written in English and did not include any textbooks or papers written in different languages.

3. Sensory Descriptive Analysis

A total of 34 studies were reviewed for sensory descriptive analysis research using special food products. Table 1 shows the categories of the samples that were used and the corresponding reference articles. Food samples could be categorized into three broad groups of (a) beverages, (b) sauces, pastes, and dressings, and (c) a group of miscellaneous 20 other specific traditional food items that were difficult to categorize. Each category is discussed in detail in the following sections.

3.1. Beverages

In the beverage category, descriptive analyses of rooibos and green tea were included. Although the tea market is growing, we considered that rooibos and green tea are still classified as ethnic foods compared to black tea. A sensory wheel for rooibos was developed by Koch et al. [19]. Those authors used a total of 69 samples evaluated by nine panelists with extensive experience on descriptive analysis. They first developed 121 descriptors during the training sessions and then 27 terms were selected for inclusion in the sensory wheel based on their relevance. After testing, 17 attributes were eventually selected for efficient sensory profiling by grouping and eliminating descriptors. The suggested application of the sensory wheel was for use for the quality control of rooibos tea. For full profiling of rooibos tea, more descriptors might be needed [18]. Jolley et al. [56] also evaluated rooibos tea; samples were compared to determine differences based on production area and harvest year. Approximately nine to 10 trained female panelists, most of whom had previous experience on rooibos evaluation, participated and a total of 208 tea samples were assessed using the 17 characteristics developed by Koch et al. [19]. Understanding the results of the study was suggested as helping to understand product segments, opening up the opportunity for marketing niche products especially at the global level.

In the case of green tea, a greater number of studies have been conducted, likely reflecting increased awareness and consumption by global consumers worldwide. Lee and Chambers [17] examined differences among green teas from different countries and the correlation with consumer data using six panelists performing a sensory descriptive analysis on six samples using 18 attributes. These findings have the potential to explain the differences among samples from different production areas, processing methods, or with different flavor characteristics. Lee and Chambers [18] developed the lexicon of green tea from nine countries (China, India, Japan, Kenya, Korea, Sri Lanka, Taiwan, Tanzania, and Vietnam) and established definitions and references. They used a total of 138 green tea samples to generate descriptors that could distinguish all kinds of green tea samples. Specifically, 31 attributes were generated by six highly trained panelists who had more than 120 h of general descriptive training and averaged more than 1200 h of sensory descriptive testing. They used the flavor profile method, which usually involves a smaller number of panelists compared to other sensory descriptive methods. The consensus procedure was used however the original flavor profile method was modified by using a 0–15 scale. Subsequently, those authors further analyzed the data to cluster green teas based on their flavor profiles and found that the origin influenced flavors through a combination of varietal differences, growing conditions, and processing variations [57]. They noted that their study could be used as a marketing tool for consumer segments for green tea so that consumers would be able to select teas that met their specific sensory preferences. Those authors also identified flavor change during storage using two green tea samples with five different storage durations [58], suggesting that the tea retailer must consider the type of packaging of green tea in order to maintain its quality during the distribution period. In addition, other research groups also worked on green tea. Lee et al. [59] developed a method to establish sample preparation and presentation procedures using six green tea samples (three green teas × two grades) and evaluating 16 characteristics. They were attempting to create a method that would minimize possible bias caused by the changes in brewing temperature that could result in differences in volatile compounds of the tea. Those authors then analyzed differences in sensory attributes such as "turbidity" or "bitter taste" between green teas that were processed under different methods and the correlation of those sensory attributes with consumer acceptability [60]. They further evaluated decaffeinated green teas as samples to ascertain the product market of decaffeinated beverages potential [61].

3.2. Sauce, Paste and Dressing

In this category, soy sauce, Eshabwe (Ghee sauce), Gochujang (Korean chili paste), soybean paste, and Danish honey were included for review. In the case of sauce or paste, different characteristics were evaluated in different countries depending on the characteristics of the product. It is likely that the foods within this category may reflect specific food cultures of each country.

Soy sauce has become a widely used sauce originating from Asian countries. Jeong et al. [62] developed a lexicon with 22 attributes for understanding and establishing a standardized descriptive analysis procedure and descriptors for fermented soy sauce in various conditions. However, that study had limited samples. Thus, the developed lexicon only had eight descriptors that were similar to those found in a later study by Cherdchu et al. [63], which was based on a larger number of samples. Those attributes were alcohol, caramel, chemical, fermented, metallic, pungent, salty, and sour. Cultural differences, sample composition, or a difference in the range of samples chosen may be the reason for the difference in the lexicon. The article by Cherdchu et al. [63] developed 58 attributes using a wide range of 20 kinds of soy sauce (selected after screening an initial set of over 120 samples). The study included the participation of panelists from Thailand and the U.S. in this cross-cultural research project. In particular, they mentioned that language and culture constituted factors that limited the ability to describe certain characteristics, although they found ways to adapt to language issues by emphasizing the importance of using standard references to conduct well-communicated evaluations in cross-cultural studies. Imamura [64] later conducted a study of 149 mostly Japanese soy sauce samples that established 88 sensory descriptors, of which many were the same descriptors as those provided by Cherdchu et al. [63]. Imamura's descriptors were divided into nine subgroups and were evaluated by 13–17 female descriptive panelists in each subgroup. A flavor wheel of soy sauce was developed to facilitate sensory evaluation and communication regarding sample qualities. In a follow-up study by Pujchakarn et al. [65], 9 female panelists, aged 38 to 56 years, developed 34 descriptors for seasoning soy sauce, a specific category of soy sauce, with the intent to provide new terms/references to add to prior soy sauce research. Soy sauce also was used as a sample for the investigation of the effects of different types of carriers [66]. That study showed changes in the flavor of soy sauce when used with differences in carriers such as rice, soup broth, and meat.

Eshabwe (Ghee sauce) comprises a traditional salty pudding-like condiment prepared from ghee or butter in western Uganda, eastern Congo, Rwanda, and Burundi [67]. Mukisa and Kiwanuka [67] established 15 characteristics with 10 panelists and evaluated the quality of samples. The developed descriptors consisted of not only flavor however also quality attributes such as soggy, musty, and stale.

They considered that standardization of the food made following traditional processes crucial. Their results may be helpful for manufacturers to standardize Eshabwe processing methods to ensure product consistency.

Gochujang (Korean chili paste) was evaluated by Kim et al. [68], who reported the development of 34 attributes by 10 panelists for 31 different samples. Their aim was to investigate the characteristics of different Gochujang made by various producers and to determine any correlation with consumer age segments differing. Gochujang dressing, a similar product, was studied by Hong et al. [69], where eight panelists developed 10 sensory terms; in addition, cross-cultural consumer tests also were conducted in Korea, China, and the U.S. The terminology of their study was used to find drivers of consumer preference and it was mentioned that an understanding of the flavors of traditional foods was important for this purpose.

Soybean paste is a well-known food in Eastern Asia. Jung et al. [20] described 18 characteristics of 14 samples for comparing data from a panel performing a sensory descriptive analysis and electronic devices such as e-nose and e-tongue. From their findings, they emphasized the importance of sensory testing for describing a complex food matrix. Kim et al. [70] examined 30 sensory attributes of various soybean paste products available in Korean markets made with different manufacturing processes and then compared these with consumer acceptability data. They indicated that consumer acceptability was significantly influenced by sweet and monosodium glutamate (MSG) flavor characteristics of products. Chung and Chung [21] previously performed cross-cultural descriptive studies from which they developed 48 terms. Notably, they mentioned that the usage pattern of the same attribute differed cross-culturally; thus, further research was suggested to overcome the linguistic differences.

In total, 27 sensory descriptors of 21 Danish honeys were created using sensory descriptive analysis [71] with the aim to describe and differentiate the uniqueness of locally produced honeys in Denmark. Ten sensory panelists, six women and four men aged 20 to 62 participated in the evaluation. Their results might be used to help communicate the specific sensory quality of honeys; however, it would be difficult to promote local uniqueness, as they did not compare the Danish honey with honeys from other countries.

3.3. Miscellaneous Products

We also identified reports on 20 food products that could not be readily categorized together (Table 1). Tofu is a typical soybean product that constitutes one of the most favored ingredients of many East Asian style cuisines [72]. Chung et al. [72] developed 27 terms for various types of tofu, concluding that the sensory characteristics of tofu varied depending on numerous factors such as brand, processing method, and ingredients. Descriptors included ones related to appearance, odor/aroma, flavor/taste, texture/mouthfeel, and aftertaste categories. Moreover, Kamizake et al. [73] showed the effect of aging of soybean under different conditions on tofu quality. The tofu samples were made using two cultivars at three different conditions of soybean aging: control, accelerated aging, and natural aging. Those authors developed 16 sensory descriptors and found that the aging of soybean affected the sensory quality of tofu such as color, flavor, and texture attributes. However, all descriptors were highly expressed in principal component 1, meaning that most terms were correlated with each other in that study. In this study, 13 panelists participated, however information regarding their age, gender, and extent of training was not included. The high correlation among attributes may suggest that the panelists were not as well trained as those in the study by Chung (72). Between the two tofu studies [72,73], odor and flavor descriptors were mostly different probably because of differences in their purpose and samples and perhaps the training and experience of panelists. However, texture attributes were comparable although different descriptors were used such as springiness, hardness, easy to cut, and stickiness [72] compared to elasticity, firmness, fracturability, and residual adherence [73], respectively.

A lexicon for baechu kimchi was developed by Chambers et al. [22]. Prior to their research, little had been published regarding this side dish/condiment despite the wide variety of available kimchi preparations and processing methods that could affect kimchi quality. Because kimchi represents a highly variable food, those authors mentioned that future research should include studies of the sensory properties of kimchi made from various ingredients. Dongchimi, another type of kimchi that is made from radish with water, was examined by Cho et al. [23]. They analyzed the relationships between sensory data from descriptive analysis and chemical properties from instrumental analysis by partial least squares regression analysis (PLSR). Further research was suggested to facilitate Dongchimi product development. Although the main ingredients differed as baechu and radish were used, respectively [22,23], the nature of kimchi utilizing similar ingredients resulted in common descriptors such as sweet, salty, sour, garlic, red pepper (chili), fermented (yeast), green, heat burn (hot), and carbonation.

Certain meat products specific to certain regions have also been evaluated using sensory descriptive analysis. Pereira et al. [24] established 14 attributes for Portuguese cooked blood sausage using 12 products. Their results may help describe the sensory qualities of the blood sausages from different producers. Blood sausage variants are found worldwide [74–76], however sensory descriptive research on blood sausage from different cultures or countries were not conducted similarly enough to compare studies. Slovenian Krvavica was evaluated to initiate protected geographical indication; however, they used a 7–point scale that was somewhat similar to a just-about-right (JAR) scale for various attributes with 4 being an optimal intensity [74]. The sensory profile of commercial Spanish dry-cured sausage was evaluated by Ruiz Pérez-Cacho et al. [77], who commented that the developed terms might define and reveal differences among samples. Marangoni and Moura [78] developed 12 sensory terms for four different treatments of Italian salami to establish their sensory profile. Jo et al. [79] determined the sensory characteristics of Bulgogi, a Korean cooked meat dish. They also conducted cross-cultural consumer acceptability testing and compared these findings to the descriptive profile to understand the preference tendencies of different cultural consumers in order to investigate the market potential abroad.

A wide range of traditional grain-based foods have been tested. Italian polenta produced with 12 different corn cultivars during 2 years of testing was studied by Zeppa et al. [80], who defined 13 descriptors for the sensory profile of traditional polenta. Nine cultivars were examined in the first year and seven in the second year. Four cultivars were examined in both years. The 13 descriptors they selected were determined to have the maximum potential to identify attributes of each sample. They indicated that the defined lexicon might be broadly applied to describe the sensory qualities of polenta, in selecting cultivars of the base ingredients, and for use in product development. Doda burfi, an Indian milk cake, from different cities was characterized in order to standardize the product for globalization and widespread marketing of traditional items by Chawla et al. [81]. In that study, a panel established 19 characteristics that could explain differences among samples. Kim et al. [82] characterized Gangjung (a traditional Korean fried puffed snack) made via different treatments using 21 attributes. They suggested that the developed descriptors might be used for both characterizing samples and explaining the effect of sample preparation.

A lexicon of turrón, a nougat product that has European protected designation of origin (PDO) status, was defined by Vázquez-Araújo et al. [25] for various brands and commercial categories. A total of 41 sensory terms were evaluated by a trained panel in the U.S. and then the terms and definitions were translated and applied for use by native Spanish turrón quality panels. Those authors indicated that generally, the lexicon was successfully implemented, however that more studies should be conducted to overcome language and cultural/scientific differences (e.g., trained panels vs. quality panelists). They noted that reference product availability could be a potential barrier for such international work because they are not always available internationally. The sensory characteristics of dates were examined by Al-Farsi et al. [83]. They developed nine sensory attributes that were compared to compositional or nutritional components to help in the promotion of dates as a healthy food.

Table 1. Sensory descriptive analysis studies using traditional foods as samples.

Food Category	Food Sample	Reference [1]	Sample Number	Descriptive Panelist Number	Number of Characteristics
Beverages	Rooibos	Jolley, Van der Rijst, Joubert, & Muller [56];	69	9	17
		Koch, Muller, Joubert, Van der Rijst, & Næs [19]	208	9–10	17
		Lee, Chambers, & Chambers [57];	138	6	32
		Lee & Chambers* [17];	6	6	18
	Green tea	Lee & Chambers [58];	10	6	20
		Lee, Lee, Kim, & Kim* [61];	8	9	13
		Lee, Lee, Sung, Lee, & Kim* [60];	7	8	15
		Lee, Chung, Lee, Lee, Kim, & Kim [59];	6	8	16
		Lee & Chambers [18]	138	6	31
Sauce paste and dressing	Soy sauce	Imamura [64];	149	13–17	88
		Pujchakarn, Suwonsichon, & Suwonsichon [65];	925	9	34
		Cherdchu & Chambers IV [66];	20	6	87–89
		Cherdchu, Chambers IV, & Suwonsichon [63];	20	9/6	58
		Jeong, Chung, Suh, Suh, & Kim [62]	6	8	22
	Eshabwe (Ghee sauce)	Mukisa & Kiwanuka [67]	4	10	15
	Gochujang (Korean chili paste)	Kim, Go, Kim, & Chung* [68]	31	10	34
	Soybean paste	Jung et al. [20];	14	8	18
		Kim, Hong, Song, Shin, & Kim* [70];	7	8	30
		Chung & Chung [21]	8	10/9	48
	Gochujang dressing	Hong, Lee, Chung, Chung, Kim, & Kim* [69]	6	8	10
	Danish honey	Stolzenbach, Byrne, & Bredie [71]	21	10	27
Miscellaneous Other Products	Tofu	Kamizake, Silva, & Prudencio* [73];	6	13	16
		Chung, Lee, & Chung [72]	7	6	27
	Kimchi	Chambers, Lee, Chun, & Miller [22];	10	5	17
		Cho, Lee, Choi, & Chung [23]	9	10	22
	Portuguese cooked blood sausage	Pereira, Dionísio, Matos, & Patarata [24]	12	18	14
	Spanish dry-cured sausage	Pérez-Cacho, Galán-Soldevilla, Crespo, & Recio [77]	7	5	26
	Italian salami	Marangoni & Moura [78]	4	12	12
	Bulgogi	Jo, Lee, Sohn, & Kim* [79]	6	8	20
	Italian polenta	Zeppa, Bertolino, & Rolle [80]	12	10	27
	Doda burfi (Indian milk cake)	Chawla, Patil, & Singh [81]	16	19	19
	Gangjung (Korean oil-puffed snack)	Kim, Kim, Chung, Lee, & Kim [82]	10	8	21
	Turrón	Vázquez-Araújo, Chambers, & Carbonell-Barrachina [25]	67	6	41
	Date	Al-Farsi, Alasalvar, Morris, Baron, & Shahidi [83]	3	10	9

[1] Asterisk (*) indicates that the study was conducted with consumer testing.

4. Consumer Studies

A total of 24 journal articles were reviewed for the consumer testing of special and unique food products. Table 2 shows the food samples that were used in each study, which were broadly categorized for the descriptive analysis as follows: beverages, sauce/paste/dressing, and miscellaneous other specific food products.

4.1. Beverages

Two different kinds of drinks accounted for products in the beverage category: green tea and açai juice. For green tea, Lee and Chambers [17] conducted consumer testing with U.S. consumers to examine how consumers liked green tea samples from three different countries and which characteristics were related to consumer preference. They analyzed their sensory descriptive and consumer data by correlation analysis and concluded that the relationship between flavor and liking was, as expected, complicated. Lee et al. [84] determined the relationship between acceptability ratings by consumers from three different countries (Korea, Thailand, and the U.S.) and flavor characteristics of green tea samples. They found that consumers who were familiar with green tea liked green vegetable attributes, however that the other group of consumers preferred green tea which did not have strong green tea flavor and bitterness. Because the acceptance of certain flavors was found to differ from one country to another (and within countries), the authors suggested that their findings implied that different food consumption experiences likely affect food selection and acceptability. Some other researchers mentioned that food consumption frequency affected acceptability [85] and familiarity influenced consumers' discrimination ability of products [86]. To investigate consumer acceptance of unfamiliar food samples, Sabbe et al. [48] used açai juices with different concentrations of fruit juice. They found that familiarity was highly associated with consumer acceptability and that further studies were required to analyze the effect of repeated exposures.

4.2. Sauce, Paste and Dressing

As with the descriptive analysis, Doenjang, Gochujang products, soy sauce, Bulgogi (sauce), and olive oils were included in this category. Doenjang is widely used in Korea to make soup, sauce, or as a condiment. To determine consumer acceptance of the product and the underlying sensory drivers, Roh et al. [46] ascertained consumer preferences and associated factors such as product form. They showed that familiar food forms usually were consumed in practice. Another study determined consumer perception of Doenjang made by different manufacturing methods [47]. Also, Kim et al. [70] classified consumers into selected groups with similar preference patterns, finding that acceptability differed according to consumer segment. To understand consumer preference and the perception of Doenjang products, they recruited over 150 consumers to participate in each study. The larger number of consumer participants was used to expand the resulting understanding of consumer preferences. Similarly, some researchers [70,87] recruited more than 200 consumers to analyze the differences of liking between different age groups. The necessary number of consumers in sensory acceptability studies was calculated as approximately 100 or more for each segment that was compared [88].

Gochujang products have also been evaluated for consumer acceptability and to better understand their acceptance. The sensory drivers of positive response in different age groups were investigated by Kim et al. [68]. Hong et al. [69] conducted cross-cultural consumer testing between Korea, China, and the U.S. to provide information on country-specific factors affecting consumer acceptance of Gochujang dressing with different ratios of each ingredient. Because the aims of both studies were to investigate consumer acceptance factors for those products, descriptive analyses were conducted first to identify sensory characteristics of the food samples. Although the samples differed in these two studies, the sweet characteristic tended to influence Korean consumers positively in both studies.

Studies also have used soy sauce and Bulgogi together as samples to evaluate consumer acceptance because soy sauce is considered a basic ingredient for Bulgogi marinade sauce. Park et al. [89]

conducted a cross-cultural study between Korean and Japanese consumer sensory perception and hedonic responses, from which they suggested that cultural background affected sensory perception. They mentioned, as other authors suggested, that cultural background affected food experiences and familiarity which could moderate consumers' sensory response to ethnic foods [53,79,90,91]. Heo and Lee [92] determined the acceptability of different brands of soy sauce and Bulgogi by U.S. consumers who were unfamiliar with the specific samples that were tested. They suggested a possible relationship between familiarity and liking because the soy sauce with familiar flavor had higher acceptability when tasted both in its original form and when used in cooking by consumers. Another study used both soy sauce and Bulgogi samples for comparing hedonic information between Korean and Chinese consumers [79]. Overall, familiarity or cultural differences appeared to influence the results of these cross-cultural studies. Korean consumers were positively influenced by the sweet taste of Bulgogi, however the opposite was true for Chinese consumers. The authors explained this finding by explaining that Chinese meat dishes usually do not have much sweet flavor. The authors implied that it was important to understand the cultural context of consumers when evaluating or promoting ethnic foods and flavors.

Olive oil has been used widely as an ingredient or sauce in Western countries. Delgado et al. [52] studied consumer liking and perceptions of extra olive virgin oils by Northern California consumers based on product packaging and labeling. That study identified some discrepancies between the overall liking evaluations based on packaging versus flavor. In their study, the "packaging group" considered the region of origin as a key point and the "blind group" evaluated the olive oil based on sensory characteristics such as bitterness or pungency. They mentioned that local or national product reputation or expectations could be an important motivation for purchasing. Pagliuca and Scarpato [51] analyzed and compared preferences of Italian olive oils between generic consumers and experts. They indicated that experts emphasized the recognition of intrinsic attributes more than novices because only the experts recognized the differences between oils from PDO compared to non-PDO oils. The degree of awareness regarding the EU certification system and geographical indications may explain this differential. Another study [93] of six olive oils compared U.S. and Spanish consumers and found that Spanish consumers liked the more bitter and green flavors of extra virgin olive oils while U.S. consumers liked the more bland, fruity, or floral flavors of the other two oils. They concluded that Spanish consumers may be more familiar with various olive oils and, thus, appreciated different characteristics. Thus, unlike most other ethnic foods, studies of olive oil have been conducted to understand consumer perceptions of various aspects in addition to their acceptance of its flavor characteristics.

Table 2. Consumer acceptability studies using traditional foods as samples.

Food Category	Food Sample	Reference [1]	Sample Number	Consumer Number	Country
Beverages	Green tea	Lee & Chambers, 2010a* [17];	12	410	USA
		Lee, Chambers IV, Chambers, Chun, Oupadissakoon, & Johnson [84]	6	120/239/100	Korea/Thailand/USA
	Açai juice	Sabbe, Verbeke, Deliza, Matta, & Van Damme [48]	6	123	Belgium
Sauce, paste and dressing	Doenjang (soybean paste)	Roh, Lee, Kim, & Kim* [46];	8	157	Korea
		Kim & Lee [47];	4	291	Korea
		Kim, Hong, Song, Shin, & Kim* [70];	7	200	Korea
	Gochujang (Korean chilli paste)	Kim, Go, Kim, & Chung* [68]	4	280	Korea
	Gochujang dressing	Hong, Lee, Chung, Chung, Kim, & Kim* [69]	6	50/34/26	Korea/China/USA [2]
	Bulgogi marinade sauces	Park, Ko, Jang, & Hong [89]	7	102/100	Korea/Japan
	Soy sauce, bulgogi	Heo & Lee, 2017 [92]	4	123	USA
		Jo, Lee, Sohn, & Kim [79]	6	76/120(72)	Korea/China
		Delgado, Gómez-Rico, & Guinard [52]	18	212	USA
	Olive oil	Pagliuca & Scarpato* [51]	5	400/35	Italy
		Vázquez-Araújo, Adhikari, Chambers, Chambers, & Carbonell-Barrachina* [93]	6	100/100	Spain/USA
Others	Idiazabal cheese	Ojeda, Etaio, Guerrero, Fernández-Gil, & Pérez-Elortondo* [42]	9	212	Spain
	Scamorza cheese	Braghieri, Piazzolla, Romaniello, Paladino, Ricciardi, & Napolitano* [43]	4	87	Italy
	Swiss cheese	Liggett, Drake, & Delwiche* [45]	15	101	USA
	Lucanian dry cured sausages	Braghieri, Piazzolla, Carlucci, Bragaglio, & Napolitano* [94]	10	102	Italy
	Argentinean Fermented Sausages	López, Bru, Vignolo, & Fadda [50]	10	120	Argentina
	Thai dried chili	Toontom, Posri, Lertsiri, & Meenune* [95]	4	120	Thailand
	Traditional Korean soup	Jang, Hong & Kim [53]	6	104/106/100	USA, Japan
	Polenta stick	Di Monaco, Miele, Volpe, Masi, & Cavella [49]	4	50	Italy
	Traditional Jordanian dessert	Saleh, Akash, Al-Dabbas, & Al-Ismail [96]	6	40	Jordan

[1] Asterisk (*) indicates that the study was conducted with sensory descriptive analysis. [2] Consumer test was conducted in Seoul, Korea, however it included consumers of each nationality.

4.3. Miscellaneous Products

The category of other foods included some cheese products, tofu, meat products, and additional foods as shown in Table 2. Although cheese comprises a well-known food product worldwide, some traditional types, which are authentic to specific regions, still constitute special and unfamiliar foods to many consumers. Ojeda et al. [42] used Idiazabal cheese with three different quality levels to compare the liking of local consumers in Spain with the sensory quality as assessed by trained panelists. In that study, the preferences of a large group of consumers was mainly driven by certain characteristics such as "sweet" and "toasty" attributes described by the trained panel. Scamorza cheese with different milk and starter type was studied by Braghieri et al. [43] in Italy who identified the driver of liking in terms of specific sensory input. They mentioned that the liking of specific sensory attributes such as appearance, taste/flavor, and texture were highly related to overall liking. They concluded that further studies were needed to promote product differentiation and to meet the sensory requirements of particular consumer segments. Swiss cheeses including different kinds of retail products were evaluated in the U.S. by Liggett et al. [45], who investigated the relationship between consumer liking and the specific flavor characteristics. For establishing the characteristics of unique food products and understanding consumer acceptance, descriptive analysis may be combined with consumer testing to better understand specific foods.

The acceptance of tofu prepared using different soybean conditions was studied by Kamizake et al. [73]. It was found that the condition of soybeans could significantly impair tofu quality.

As with descriptive analysis, certain meat products of specific regions were used for consumer studies. Lucanian dry cured sausages were examined by Braghieri et al. [94] who found that both intrinsic and extrinsic aspects such as taste, animal health, and preservation affected consumer choice. López et al. [50] explained that consumer acceptance of Argentinean fermented sausages from different regions represented a quality factor that was as important as instrumental analysis to understand differences. These results suggested that consumer acceptability constitutes an important component of quality control.

The sensory characteristics of Thai dried chili prepared with different treatments were determined by Toontom et al. [95] along with consumer testing. Familiarity and positive experiences were suggested to be associated with consumer acceptance. However, to understand international consumer responses and target the global market, cross-cultural consumer studies are needed. To this end, Jang et al. [53] compared the consumer acceptance of different kinds of Korean traditional soup between U.S. and Japanese consumers. They suggested that strategic modification of food flavor might help decrease neophobic responses from target consumers. In turn, the dynamic overall preference of polenta sticks was studied by Di Monaco et al. [49], which indicated that both food properties and food consumption duration affected the preference for a product. Finally, Saleh et al. [96] examined the effect of filling on the acceptability of a Jordanian traditional dessert, which determined that the ratio of ingredients could affect product acceptability for consumers. In their study, the filling with a high date ratio mostly affected hardness of the sample and its acceptability.

5. Discussion

Numerous kinds of unique and special food products exist, however only a small number of those have been studied using good sensory methods. We summarized the sensory studies of traditional and authentic food products. There were research projects conducted for both sensory descriptive analysis and consumer study to develop attributes, analyze consumer preference/perception, and identify the factors affecting consumer acceptance. Both methods were important to understand food products and their consumer perception. Where only sensory descriptive analysis was conducted, researchers could understand the attributes and intensities of the sample and potentially differentiate among samples with different ingredients, processing, storage conditions, etc. However, researchers could not determine which attributes would be related to consumer acceptance. Similarly, studies that used only consumers could understand consumer preference, however they could not prove which attributes affected consumer acceptance and perception. For the deepest understanding of consumer products, both methods should be used.

Some limitations were observed in the studies. For descriptive analysis in this review, there were large variations in the numbers of panelists that were used. The minimum panelist number was 5 [22] and the maximum number of panelists was 18–19 [24,81]. It is not certain whether the difference in the number of panelists affects the reliability of the sensory results, however it is clear that there are differences between methods for each descriptive study. Actually, the recommended ideal number of judges was not clear. Although the number of panelists varies by method, researchers should provide a clear basis if their participant number was sufficient. Chambers et al. [97] pointed out that the number of panelists necessary for descriptive panels was dependent on the training of the assessors and the types of attributes that were evaluated, with fewer being needed when the training/experience level was higher or when the attributes were more easily evaluated.

There was also variation among the number of participants in consumer tests. Consumers who participated in consumer tests should be representative of potential consumers of the product [97]. Thus, generally 100 or more consumers need to be recruited per segment for quantitative consumer tests to be reliable. Because some of the consumer tests in this review were conducted with small numbers of consumers, it is important to understand that the reliability of those studies may be less than others. For example, Hong et al. [69] conducted a cross-cultural study of Gochujang dressing with Korean (n = 50), Chinese (n = 34), and U.S. (n = 26) citizens. Although the total number of consumers was more than 100, participants from each country were not enough to be representative of populations in each country. Other research [49,96] also had small numbers of consumers. Consumer tests should be conducted with sufficient numbers of consumers because otherwise they may present unreliable results that are not representative of the population. If conducting consumer studies with a smaller number of participants, researchers must justify their reasoning for deviating from recommendations.

6. Conclusions

In this review, we reviewed sensory analysis (i.e., evaluation) studies regarding traditional and authentic food products, which sometimes are called specialty, exotic, or unique foods. The purpose of our review was to provide a broad summary of the descriptive analyses and consumer research that is currently available for various traditional and authentic food products. As globalization progresses, the need for understanding flavors of traditional, unique, regional, or artisanal foods is clearly increasing in order to better comprehend and accommodate consumer demands for unfamiliar food products. Additional sensory science research using unusual and ethnic food products should be carried out to better understand product characteristics.

Numerous studies highlighted the effect of familiarity on consumer perception. In particular, a positive correlation between familiarity and consumer acceptance was reported frequently [91,97–100]. The consumers who were familiar with the sample showed higher liking than non-familiar consumers [91]. In addition, descriptive analysis could also be affected by familiarity. Drake et al. [101] noted that lexicon development could not be freed from the influence of the culture from which the panel originated, although other authors [25,63] have reported that cross-cultural studies using lexicons certainly are possible with planning and coordination.

To obtain deeper insights regarding specialty and unique food products, large numbers of studies using different kinds of traditional, unique, or ethnic food items and categories from a range of countries or regions or made from various ingredients or processing methods should be carried out. In addition, it is necessary to conduct cross-cultural descriptive and consumer studies to establish universal lexicons and to understand the attributes that influence consumers' liking.

Author Contributions: J.Y. helped to write this article and J.L. conceived, guided, and revised this article.

Funding: This work was supported by a 2-Year Research Grant of Pusan National University.

Conflicts of Interest: The authors declare no conflict of interest.

References

1. Tuorila, H.; Monteleone, E. Sensory food science in the changing society: Opportunities, needs, and challenges. *Trends Food Sci. Technol.* **2009**, *20*, 54–62. [CrossRef]
2. Peryam, D.R.; Pilgrim, F.J. Hedonic scale method of measuring food preferences. *Food Technol.* **1957**, *11*, 9–14.
3. Lawless, H.T.; Heymann, H. Introduction. In *Sensory Evaluation of Food: Principles and Practices*, 2nd ed.; Lawless, H.T., Heymann, H., Eds.; Springer Science and Business Media: New York, NY, USA, 2010; pp. 1–2, ISBN 978-1-4419-6487-8.
4. Meilgaard, M.M.; Civille, G.V.; Carr, B.T. Descriptive Analysis Techniques. In *Sensory Evaluation Techniques*, 5th ed.; Meilgaard, M.M., Civille, G.V., Carr, B.T., Eds.; CRC Press: Boca Raton, FL, USA, 2016; pp. 201–219, ISBN 978-1-4822-1690-5.
5. Murray, J.M.; Delahunty, C.M.; Baxter, I.A. Descriptive sensory analysis: Past, present and future. *Food Res. Int.* **2001**, *34*, 461–471. [CrossRef]
6. Cairncross, S.E.; Sjostrom, L.B. Flavor profiles—A new approach to flavor problems. *Food Technol.* **1950**, *4*, 308–311. [CrossRef]
7. Civille, G.V.; Szczesniak, A.S. Guidelines to training a texture profile panel. *J. Texture Stud.* **1973**, *4*, 204–223. [CrossRef]
8. Stone, H.; Sidel, J.; Oliver, S.; Woolsey, A.; Singleton, R.C. Sensory evaluation by quantitative descriptive analysis. In *Descriptive Sensory Analysis in Practice*; Gacular, M.C., Jr., Ed.; Food & Nutrition Press, Inc.: Trumbull, CT, USA, 2004; pp. 23–34, ISBN 9780917678370.
9. Venturi, F.; Sanmartin, C.; Taglieri, I.; Nari, A.; Andrich, G.; Zinnai, A. Effect of the baking process on artisanal sourdough bread-making: A technological and sensory evaluation. *Agrochimica* **2016**, *60*, 222–234. [CrossRef]
10. Lestringant, P.; Delarue, J.; Heymann, H. 2010-2015: How have conventional descriptive analysis methods really been used? A systematic review of publications. *Food Qual. Preference* **2018**, *71*, 1–7. [CrossRef]
11. D'Antuono, L.F.; Bignami, C. Perception of typical Ukrainian foods among an Italian population. *Food Qual. Preference* **2012**, *25*, 1–8. [CrossRef]
12. Groves, A.M. Authentic British food products: A review of consumer perceptions. *Int. J. Consumer Stud.* **2001**, *25*, 246–254. [CrossRef]
13. Jang, S.S.; Ha, J.; Park, K. Effects of ethnic authenticity: Investigating Korean restaurant customers in the US. *Int. J. Hosp. Manag.* **2012**, *31*, 990–1003. [CrossRef]
14. Bäckström, A.; Pirttilä-Backman, A.M.; Tuorila, H. Dimensions of novelty: A social representation approach to new foods. *Appetite* **2003**, *40*, 299–307. [CrossRef]
15. Kwon, D.Y. What is ethnic food? *J. Ethn. Foods* **2015**, *2*, 1. [CrossRef]
16. Dwyer, J.; Bermudez, O.I. Ethnic Foods. In *Encyclopedia of Food Science and Nutrition*, 2nd ed.; Caballero, B., Finglas, P., Toldra, F., Eds.; Academic Press: Cambridge, MA, USA, 2003; pp. 2190–2195, ISBN 978-0-12-227055-0.
17. Lee, J.; Chambers, D.H. Descriptive analysis and US consumer acceptability of 6 green tea samples from China, Japan, and Korea. *J. Food Sci.* **2010**, *75*, 141–147. [CrossRef]
18. Lee, J.; Chambers, D.H. A lexicon for flavor descriptive analysis of green tea. *J. Sensory Stud.* **2007**, *22*, 256–272. [CrossRef]
19. Koch, I.S.; Muller, M.; Joubert, E.; van der Rijst, M.; Næs, T. Sensory characterization of rooibos tea and the development of a rooibos sensory wheel and lexicon. *Food Res. Int.* **2012**, *46*, 217–228. [CrossRef]
20. Jung, H.Y.; Kwak, H.S.; Kim, M.J.; Kim, Y.; Kim, K.O.; Kim, S.S. Comparison of a descriptive analysis and instrumental measurements (electronic nose and electronic tongue) for the sensory profiling of Korean fermented soybean paste (doenjang). *J. Sensory Stud.* **2017**, *32*, e12282. [CrossRef]
21. Chung, L.; Chung, S.J. Cross-cultural comparisons among the sensory characteristics of fermented soybean using Korean and Japanese descriptive analysis panels. *J. Food Sci.* **2007**, *72*, 676–688. [CrossRef] [PubMed]
22. Chambers, E., IV; Lee, J.; Chun, S.; Miller, A.E. Development of a lexicon for commercially available cabbage (baechu) kimchi. *J. Sensory Stud.* **2012**, *27*, 511–518. [CrossRef]
23. Cho, J.H.; Lee, S.J.; Choi, J.J.; Chung, C.H. Chemical and sensory profiles of dongchimi (Korean watery radish kimchi) liquids based on descriptive and chemical analyses. *Food Sci. Biotechnol.* **2015**, *24*, 497–506. [CrossRef]

24. Pereira, J.A.; Dionísio, L.; Matos, T.J.S.; Patarata, L. Sensory lexicon development for a Portuguese cooked blood sausage–Morcela de Arroz de Monchique–to predict its usefulness for a geographical certification. *J. Sensory Stud.* **2015**, *30*, 56–67. [CrossRef]
25. Vázquez-Araújo, L.; Chambers, D.; Carbonell-Barrachina, Á.A. Development of a sensory lexicon and application by an industry trade panel for turrón, a European protected product. *J. Sensory Stud.* **2012**, *27*, 26–36. [CrossRef]
26. Drake, M.A. Invited review: Sensory analysis of dairy foods. *J. Dairy Sci.* **2007**, *90*, 4925–4937. [CrossRef] [PubMed]
27. Venturi, F.; Andrich, G.; Sanmartin, C.; Scalabrelli, G.; Ferroni, G.; Zinnai, A. The expression of a full-bodied red wine as a function of the characteristics of the glass utilized for the tasting. *CyTA-J. Food.* **2014**, *12*, 291–297. [CrossRef]
28. Villagas, B.; Carbonell, I.; Costell, E. Acceptability of milk and soymilk vanilla beverages: Demographics consumption frequency and sensory aspects. *Food Sci. Technol. Int.* **2009**, *15*, 203–210. [CrossRef]
29. Jaeger, S. Non-sensory factors in sensory science research. *Food Qual. Preference* **2006**, *17*, 132–144. [CrossRef]
30. Urala, N.; Lähteenmäki, L. Attitudes behind consumers' willingness to use functional foods. *Food Qual. Preference* **2004**, *15*, 793–803. [CrossRef]
31. Köster, E.P. Diversity in the determinants of food choice: A psychological perspective. *Food Qual. Preference* **2009**, *20*, 70–82. [CrossRef]
32. Rozin, P. Cultural approaches to human food preferences. In *Nutritional Modulation of Neural Function*; Morley, J.E., Sterman, M.B., Walsh, J.H., Eds.; Academic Press: San Diego, CA, USA, 1988; pp. 137–153, ISBN 978-0-12-506455-2.
33. Prescott, J.; Bell, G. Cross-cultural determinants of food acceptability: Recent research on sensory perceptions and preferences. *Trends Food Sci. Technol.* **1995**, *6*, 201–205. [CrossRef]
34. Frez-Muñoz, L.; Steenbekkers, B.L.; Fogliano, V. The choice of canned whole peeled tomatoes is driven by different key quality attributes perceived by consumers having different familiarity with the product. *J. Food Sci.* **2016**, *81*, 2988–2996. [CrossRef]
35. Raudenbush, B.; Frank, R.A. Assessing food neophobia: The role of stimulus familiarity. *Appetite* **1999**, *32*, 261–271. [CrossRef]
36. Ares, G. Methodological issues in cross-cultural sensory and consumer research. *Food Qual. Preference* **2018**, *64*, 253–263. [CrossRef]
37. Hu, X.; Lee, J. Emotions elicited while drinking coffee: A cross-cultural comparison between Korean and Chinese consumers. *Food Qual. Preference* **2018**. [CrossRef]
38. Meiselman, H.L. A review of the current state of emotion research in product development. *Food Res. Int.* **2015**, *76*, 192–199. [CrossRef]
39. Kim, M.K.; Lee, Y.J.; Kwak, H.S.; Kang, M.W. Identification of sensory attributes that drive consumer liking of commercial orange juice products in Korea. *J. Food Sci.* **2013**, *78*, S1451–S1458. [CrossRef] [PubMed]
40. Bayarri, S.; Marti, M.; Carbonell, I.; Costell, E. Identifying drivers of liking for commercial spreadable cheeses with different fat content. *J. Sensory Stud.* **2012**, *27*, 1–11. [CrossRef]
41. King, S.C.; Meiselman, H.L. Development of a method to measure consumer emotions associated with foods. *Food Qual. Preference* **2010**, *21*, 168–177. [CrossRef]
42. Ojeda, M.; Etaio, I.; Guerrero, L.; Fernández-Gil, M.P.; Pérez-Elortondo, F.J. Does consumer liking fit the sensory quality assessed by trained panelists in traditional food products? A study on PDO Idiazabal cheese. *J. Sensory Stud.* **2018**, *33*, e12318. [CrossRef]
43. Braghieri, A.; Piazzolla, N.; Romaniello, R.; Paladino, F.; Ricciardi, A.; Napolitano, F. Effect of adjuncts on sensory properties and consumer liking of Scamorza cheese. *J. Dairy Sci.* **2015**, *98*, 1479–1491. [CrossRef] [PubMed]
44. Braghieri, A.; Girolami, A.; Riviezzi, A.M.; Piazzolla, N.; Napolitano, F. Liking of traditional cheese and consumer willingness to pay. *Italian J. Anim. Sci.* **2014**, *13*, 3029. [CrossRef]
45. Liggett, R.E.; Drake, M.A.; Delwiche, J.F. Impact of flavor attributes on consumer liking of Swiss cheese. *J. Dairy Sci.* **2008**, *91*, 466–476. [CrossRef] [PubMed]
46. Roh, S.H.; Lee, S.M.; Kim, S.S.; Kim, K.O. Importance of applying condiments in a commonly consumed food system for understanding the association between familiarity and sensory drivers of liking: A study focused on Doenjang. *J. Food Sci.* **2018**, *83*, 464–474. [CrossRef] [PubMed]

47. Kim, M.K.; Lee, K.G. Correlating consumer perception and consumer acceptability of traditional Doenjang in Korea. *J. Food Sci.* **2014**, *79*, S2330–S2336. [CrossRef] [PubMed]
48. Sabbe, S.; Verbeke, W.; Deliza, R.; Matta, V.M.; Van Damme, P. Consumer liking of fruit juices with different açaí (*Euterpe oleracea* Mart.) concentrations. *J. Food Sci.* **2009**, *74*, 171–176. [CrossRef] [PubMed]
49. Di Monaco, R.; Miele, N.A.; Volpe, S.; Masi, P.; Cavella, S. Temporal dominance of sensations and dynamic liking evaluation of polenta sticks. *Br. Food J.* **2016**, *118*, 749–760. [CrossRef]
50. López, C.M.; Bru, E.; Vignolo, G.M.; Fadda, S.G. Main factors affecting the consumer acceptance of Argentinean fermented sausages. *J. Sensory Stud.* **2012**, *27*, 304–313. [CrossRef]
51. Pagliuca, M.M.; Scarpato, D. The olive oil sector: A comparison between consumers and "experts" choices by the sensory analysis. *Proc. Econ. Financ.* **2014**, *17*, 221–230. [CrossRef]
52. Delgado, C.; Gómez-Rico, A.; Guinard, J.X. Evaluating bottles and labels versus tasting the oils blind: Effects of packaging and labeling on consumer preferences, purchase intentions and expectations for extra virgin olive oil. *Food Res. Int.* **2013**, *54*, 2112–2121. [CrossRef]
53. Jang, S.H.; Hong, J.H.; Kim, M.Y. Consumer acceptability and purchase intent of traditional Korean soup in the United States and Japan. *Food Sci. Biotechnol.* **2014**, *23*, 389–400. [CrossRef]
54. Agriculture and Agri-Food Canada. Global Consumer Trends: Sensory Food Experiences. Market Analysis Report; September 2011. Available online: http://www.cme-mec.ca/_uploads/_media/gt64jqkt.pdf (accessed on 3 October 2018).
55. Hsu, A. U.S. Is A Spicier Nation (Literally) Since 1970s. Available online: https://www.npr.org/templates/story/story.php?storyId=128852866 (accessed on 3 October 2018).
56. Jolley, B.; Van der Rijst, M.; Joubert, E.; Muller, M. Sensory profile of rooibos originating from the Western and Northern Cape governed by production year and development of rooibos aroma wheel. *S. Afr. J. Bot.* **2017**, *110*, 161–166. [CrossRef]
57. Lee, J.; Chambers, D.H.; Chambers, E., 4th. A comparison of the flavor of green teas from around the world. *J. Sci. Food Agric.* **2014**, *94*, 1315–1324. [CrossRef]
58. Lee, J.; Chambers, D.H. Flavors of green tea change little during storage. *J. Sensory Stud.* **2010**, *25*, 512–520. [CrossRef]
59. Lee, S.M.; Chung, S.J.; Lee, O.H.; Lee, H.S.; Kim, Y.K.; Kim, K.O. Development of sample preparation, presentation procedure and sensory descriptive analysis of green tea. *J. Sensory Stud.* **2008**, *23*, 450–467. [CrossRef]
60. Lee, O.H.; Lee, H.S.; Sung, Y.E.; Lee, S.M.; Kim, K.O. Sensory characteristics and consumer acceptability of various green teas. *Food Sci. Biotechnol.* **2008**, *17*, 349–356.
61. Lee, S.M.; Lee, H.S.; Kim, K.H.; Kim, K.O. Sensory characteristics and consumer acceptability of decaffeinated green teas. *J. Food Sci.* **2009**, *74*, 135–141. [CrossRef] [PubMed]
62. Jeong, S.Y.; Chung, S.J.; Suh, D.S.; Suh, B.C.; Kim, K.O. Developing a descriptive analysis procedure for evaluating the sensory characteristics of soy sauce. *J. Food Sci.* **2004**, *69*, 319–325. [CrossRef]
63. Cherdchu, P.; Chambers, E., IV; Suwonsichon, T. Sensory lexicon development using trained panelists in Thailand and the USA: Soy sauce. *J. Sensory Stud.* **2013**, *28*, 248–255. [CrossRef]
64. Imamura, M. Descriptive terminology for the sensory evaluation of soy sauce. *J. Sensory Stud.* **2016**, *31*, 393–407. [CrossRef]
65. Pujchakarn, T.; Suwonsichon, S.; Suwonsichon, T. Development of a sensory lexicon for a specific subcategory of soy sauce: Seasoning soy sauce. *J. Sensory Stud.* **2016**, *31*, 443–452. [CrossRef]
66. Cherdchu, P.; Chambers, E., IV. Effect of carriers on descriptive sensory characteristics: A case study with soy sauce. *J. Sensory Stud.* **2014**, *29*, 272–284. [CrossRef]
67. Mukisa, I.M.; Kiwanuka, B.J. Traditional processing, composition, microbial quality and sensory characteristics of Eshabwe (ghee sauce). *Int. J. Dairy Technol.* **2018**, *71*, 149–157. [CrossRef]
68. Kim, M.R.; Go, J.E.; Kim, H.Y.; Chung, S.J. Understanding the sensory characteristics and drivers of liking for gochujang (Korean fermented chili pepper paste). *Food Sci. Biotechnol.* **2017**, *26*, 409–418. [CrossRef] [PubMed]
69. Hong, J.H.; Lee, K.W.; Chung, S.; Chung, L.; Kim, H.R.; Kim, K.O. Sensory characteristics and cross-cultural comparisons of consumer acceptability for Gochujang dressing. *Food Sci. Biotechnol.* **2012**, *21*, 829–837. [CrossRef]

70. Kim, H.G.; Hong, J.H.; Song, C.K.; Shin, H.W.; Kim, K.O. Sensory characteristics and consumer acceptability of fermented soybean paste (Doenjang). *J. Food Sci.* **2010**, *75*, S375–S383. [CrossRef] [PubMed]
71. Stolzenbach, S.; Byrne, D.V.; Bredie, W.L.P. Sensory local uniqueness of Danish honeys. *Food Res. Int.* **2011**, *44*, 2766–2774. [CrossRef]
72. Chung, J.A.; Lee, H.S.; Chung, S.J. Developing sensory lexicons for tofu. *Food Qual. Culture* **2008**, *2*, 27–31.
73. Kamizake, N.K.K.; Silva, L.C.P.; Prudencio, S.H. Impact of soybean aging conditions on tofu sensory characteristics and acceptance. *J. Sci. Food Agric.* **2018**, *98*, 1132–1139. [CrossRef] [PubMed]
74. Gašperlin, L.; Skvarča, M.; Žlender, B.; Lušnic, M.; Polak, T. Quality Assessment of Slovenian Krvavica, A Traditional Blood Sausage: Sensory Evaluation. *J. Food Process. Preserv.* **2014**, *38*, 97–105. [CrossRef]
75. Choi, Y.S.; Choi, J.H.; Han, D.J.; Kim, H.Y.; Lee, M.A.; Kim, H.W.; Lee, C.H.; Paik, H.D.; Kim, C.J. Physicochemical and sensory characterization of Korean blood sausage with added rice bran fiber. *Korean J. Food Sci. Anim. Resour.* **2009**, *29*, 260–268. [CrossRef]
76. Diez, A.M.; Santos, E.M.; Jaime, I.; Rovira, J. Application of organic acid salts and high-pressure treatments to improve the preservation of blood sausage. *Food Microbiol.* **2008**, *25*, 154–161. [CrossRef]
77. Ruiz Pérez-Cacho, M.P.; Galán-Soldevilla, H.; León Crespo, F.; Molina Recio, G. Determination of the sensory attributes of a Spanish dry-cured sausage. *Meat Sci.* **2005**, *71*, 620–633. [CrossRef]
78. Marangoni, C.; de Moura, N.F. Sensory profile of Italian salami with coriander (*Coriandrum sativum* L.) essential oil. *Food Sci. Technol.* **2011**, *31*, 119–123. [CrossRef]
79. Jo, S.G.; Lee, S.M.; Sohn, K.H.; Kim, K.O. Sensory characteristics and cross-cultural acceptability of Chinese and Korean consumers for ready-to-heat (RTH) type bulgogi (Korean traditional barbecued beef). *Food Sci. Biotechnol.* **2015**, *24*, 921–930. [CrossRef]
80. Zeppa, G.; Bertolino, M.; Rolle, L. Quantitative descriptive analysis of Italian polenta produced with different corn cultivars. *J. Sci. Food Agric.* **2012**, *92*, 412–417. [CrossRef] [PubMed]
81. Chawla, R.; Patil, G.R.; Singh, A.K. Sensory characterization of doda burfi (Indian milk cake) using Principal Component Analysis. *J. Food Sci. Technol.* **2014**, *51*, 558–564. [CrossRef] [PubMed]
82. Kim, H.R.; Kim, K.M.; Chung, S.J.; Lee, J.W.; Kim, K.O. Effects of steeping conditions of waxy rice on the physical and sensory characteristics of Gangjung (a traditional Korean oil-puffed snack). *J. Food Sci.* **2007**, *72*, 544–550. [CrossRef]
83. Al-Farsi, M.; Alasalvar, C.; Morris, A.; Baron, M.; Shahidi, F. Compositional and sensory characteristics of three native sun-dried date (*Phoenix dactylifera* L.) varieties grown in Oman. *J. Agric. Food Chem.* **2005**, *53*, 7586–7591. [CrossRef] [PubMed]
84. Lee, J.; Chambers, E., IV; Chambers, D.H.; Chun, S.S.; Oupadissakoon, C.; Johnson, D.E. Consumer acceptance for green tea by consumers in the United States, Korea and Thailand. *J. Sensory Stud.* **2010**, *25*, 109–132. [CrossRef]
85. Borgogno, M.; Favotto, S.; Corazzin, M.; Cardello, A.V.; Piasentier, E. The role of product familiarity and consumer involvement on liking and perceptions of fresh meat. *Food Qual. Preference* **2015**, *44*, 139–147. [CrossRef]
86. Choi, Y.; Lee, J. The effect of extrinsic cues on consumer perception: A study using milk tea products. *Food Qual. Preference* **2019**, *71*, 343–353. [CrossRef]
87. Kwak, H.S.; Jung, H.Y.; Kim, M.J.; Kim, S.S. Differences in consumer perception of Korean traditional soybean paste (Doenjang) between younger and older consumers by blind and informed tests. *J. Sensory Stud.* **2017**, *32*, e12302. [CrossRef]
88. Hough, G.; Wakeling, I.; Mucci, A.; Chambers, E., IV; Gallardo, I.M.; Alves, L.R. Number of consumers necessary for sensory acceptability tests. *Food Qual. Preference* **2006**, *17*, 522–526. [CrossRef]
89. Park, H.J.; Ko, J.M.; Jang, S.H.; Hong, J.H. Comparison of consumer perception and liking of bulgogi marinade sauces between Korea and Japan using flash profiling. *Food Sci. Biotechnol.* **2017**, *26*, 427–434. [CrossRef] [PubMed]
90. Jang, S.H.; Kim, M.J.; Lim, J.; Hong, J.H. Cross-cultural comparison of consumer acceptability of kimchi with different degree of fermentation. *J. Sensory Stud.* **2016**, *31*, 124–134. [CrossRef]
91. Hong, J.H.; Park, H.S.; Chung, S.J.; Chung, L.; Cha, S.M.; Lê, S.; Kim, K.O. Effect of Familiarity on a Cross-Cultural Acceptance of a Sweet Ethnic Food: A Case Study with Korean Traditional Cookie (Yackwa). *J. Sensory Stud.* **2014**, *29*, 110–125. [CrossRef]

92. Heo, J.; Lee, J. US consumers' acceptability of soy sauce and *bulgogi*. *Food Sci. Biotechnol.* **2017**, *26*, 1271–1279. [CrossRef] [PubMed]
93. Vázquez-Araújo, L.; Adhikari, K.; Chambers, E., IV; Chambers, D.; Carbonell-Barrachina, A. Cross-cultural perception of six commercial olive oils: A study with Spanish and US consumers. *Food Sci. Technol. Int.* **2015**, *21*, 454–466.
94. Braghieri, A.; Piazzolla, N.; Carlucci, A.; Bragaglio, A.; Napolitano, F. Sensory properties, consumer liking and choice determinants of Lucanian dry cured sausages. *Meat Sci.* **2016**, *111*, 122–129. [CrossRef]
95. Toontom, N.; Posri, W.; Lertsiri, S.; Meenune, M. Effect of drying methods on Thai dried chilli's hotness and pungent odour characteristics and consumer liking. *Int. Food Res. J.* **2016**, *23*, 289–299.
96. Saleh, M.; Akash, M.W.; Al-Dabbas, M.; Al-Ismail, K.M. Sesame-oil-cake (SOC) impacted consumer liking of a traditional Jordanian dessert; a mixture response surface model approach. *Life Sci. J.* **2014**, *11*, 38–44.
97. Chambers, E., IV; Bowers, J.A.; Dayton, A.D. Statistical designs and panel training/experience for sensory analysis. *J. Food Sci.* **1981**, *46*, 1902–1906. [CrossRef]
98. Tuorila, H.; Huotilainen, A.; Lähteenmäki, L.; Ollila, S.; Tuomi-Nurmi, S.; Urala, N. Comparison of affective rating scales and their relationship to variables reflecting food consumption. *Food Qual. Preference* **2008**, *19*, 51–61. [CrossRef]
99. Lawless, L.J.; Threlfall, R.T.; Meullenet, J.F.; Howard, L.R. Applying a mixture design for consumer optimization of black cherry, Concord grape and pomegranate juice blends. *J. Sensory Stud.* **2013**, *28*, 102–112. [CrossRef]
100. Lawless, L.J.; Threlfall, R.T.; Meullenet, J.F. Using a choice design to screen nutraceutical-rich juices. *J. Sensory Stud.* **2013**, *28*, 113–124. [CrossRef]
101. Drake, M.A.; Yates, M.D.; Gerard, P.D.; Delahunty, C.M.; Sheehan, E.M.; Turnbull, R.P.; Dodds, T.M. Comparison of differences between lexicons for descriptive analysis of Cheddar cheese flavour in Ireland, New Zealand, and the United States of America. *Int. Dairy J.* **2005**, *15*, 473–483. [CrossRef]

© 2019 by the authors. Licensee MDPI, Basel, Switzerland. This article is an open access article distributed under the terms and conditions of the Creative Commons Attribution (CC BY) license (http://creativecommons.org/licenses/by/4.0/).

Article

Influence of Monosodium Glutamate and Its Substitutes on Sensory Characteristics and Consumer Perceptions of Chicken Soup

Shangci Wang, Shaokang Zhang and Koushik Adhikari *

Department of Food Science and Technology, University of Georgia-Griffin Campus, 1109 Experiment Street, Griffin, GA 30223, USA; wsc727@uga.edu (S.W.); zskzsk@uga.edu (S.Z.)
* Correspondence: koushik7@uga.edu; Tel.: +1-770-412-4736

Received: 16 December 2018; Accepted: 12 February 2019; Published: 14 February 2019

Abstract: Soup manufacturers are removing monosodium glutamate (MSG) to meet consumer demand for natural ingredients. This research investigated the influence of MSG and its substitutes (yeast extract: YE; mushroom concentrate: MC; tomato concentrate: TC) on clear chicken soup with 0.4% sodium chloride (salt) by comparing sensory attributes and consumer acceptability among each other, and to a chicken soup sample containing 0.5% salt (Salt 0.5%). The soup with 0.4% salt without enhancers was designated as the control. Corresponding list of ingredients was also presented to consumers to study the effects on consumer expectations about chicken soup products. Our results showed that MSG and its substitutes significantly ($P < 0.05$) enhanced the sensory properties of chicken soup. These flavor enhancers also achieved statistically same or stronger improvement in overall flavor, meaty flavor, chicken flavor and umami taste when compared to Salt 0.5% sample. Consumers significantly preferred MSG 0.1%, YE 0.025%, and Salt 0.5% samples than others. Compared to MC and TC samples, less consumers perceived MSG and YE samples as "free of artificial" and "natural" with lower consumption interest. Claims about artificial/natural ingredients were attractive selling points for chicken soups, but good sensory appealing was the most important attribute linearly affecting consumer satisfactions.

Keywords: monosodium glutamate (MSG); MSG substitutes; food label; chicken soup

---

1. Introduction

Soup is an important food product category in the market. According to Grand View Research, Inc., the US soup market size is expected to reach ~7.7 billion US dollars by 2025 [1]. Chicken soup has been known for centuries as home remedy for cold and flu, possibly due to its hot temperature and anti-inflammatory effects [2,3]. In addition, chicken soup can help with weight control and reduction because of its high water content. There are several trends that drive the growth of soup industry, including demands for natural and fresh ingredients and rising awareness of healthy food choice. To further fit the natural and healthy portfolio, several major soup manufacturers have announced to move away from using artificial ingredients and flavors in their products. Monosodium glutamate (MSG) is one such ingredient that has been controversial for decades. It is one of the ingredients that some companies have committed to remove from products.

MSG is a flavor enhancer commonly added to processed food products like chicken soup to boost the palatability. Its remarkable effects on the sensory appeal have been proven in various studies [4–6]. Removal of this ingredient is very likely to cause a reduced consumer acceptability. Using MSG substitute is a promising approach to compensate for the sensory satisfaction loss caused by MSG elimination. The flavor enhancement effect of MSG is mainly from glutamate which contributes to umami or savory taste sensation. Besides glutamate, there are several other umami eliciting components such as aspartate and 5′-ribonicleotides. Among nucleotides, inosinate (IMP) and guanylate (GMP) significantly contribute to flavor and taste enhancement. Theoretically, substances that are naturally rich in umami components have the potential to replace MSG in food products. Wang and Adhikari found that consumers preferred natural extracts such as yeast extract, mushroom extract, and tomato extract as MSG substitute in chicken soup products [7]. Yeast typically has around 7–12% of RNA content which makes it a good source to produce nucleotide-rich ingredients [8]. Previous studies found that compared to MSG, yeast extract (YE) had stronger salty and umami taste at the same usage level and better improved the tastes of meat samples [5,9]. Mushroom is another nucleotide-rich material. For instance, GMP is originally isolated from shiitake mushroom. Ground mushroom can substitute 80% beef in taco blend with enhanced overall flavor and umami taste [10]. Tomato is very rich in glutamate. In contrast to MSG, tomato puree can better enhance the sweet, salty, and sour taste of minced beef but suppress the beefy flavor [5]. In addition, processing like drying can remarkably boost the glutamate content in tomato and mushroom and generate more GMP in mushroom [11]. Currently, there is limited research comparing the enhancement effects of MSG with these natural extracts in food products.

Excessive sodium intake is major health concern in the US, which makes sodium reduction almost a necessity in soup manufacturing. Unfortunately, reducing table salt usage always results in consumer complaints of bland taste and flavor. Application of MSG is a common approach to improve palatability of low sodium soups. Given the capability of salty taste enhancement, MSG substitute may also be able to increase the sensory appeal of soup with reduced salt content. Previous study indicated that 1% and 2% yeast extract successfully enhanced the taste of fermented sausage that had 25% NaCl replaced by KCl [12]. Ground mushroom has also been reported to improve the flavor of taco blend with 25% sodium reduction but failed to mitigate the reduced salty taste [10]. To replace MSG, it is necessary to conduct more research to compare the performance between MSG and its alternatives in salt-reduced food matrix.

Consumer's food choice is a tradeoff among sensory and non-sensory factors [13]. As the rising health awareness, consumers use food labels to obtain the information they require. Information on food packaging plays a crucial role in consumer purchasing decisions. Consumers are likely to imagine the tastes and expect the health benefits of the product based on the important cues from packaging, which might further influence their hedonic and sensory perceptions [14]. Bandara et al. reported that MSG content affected the purchase intent of 21% of the respondents to a large extent [15]. However, Prescott and Young noted that information about MSG usage did not influence consumer liking and natural taste perception about chicken soups [16]. Thus, it would be interesting to investigate how the information about MSG and its substitute usage affects consumer expectations and perceptions about chicken soup products. Besides ingredient choice, there are several other claims that commonly appear on packaging of chicken soup products. However, it is unclear how these features influence consumer satisfactions. Kano analysis is a useful consumer research tool that is widely practiced in industries to classify and prioritize product features based on their influences on customer satisfaction.

This work applied MSG and its substitutes in chicken soup added with 0.4% NaCl to (1) compare their influences on sensory attributes and consumer acceptability of chicken soup and (2) investigate whether samples added with these flavor enhancers could achieve the sensory appeal of the chicken soup added with 0.5% salt. This study also aimed to analyze how information about MSG and its replacement usage affects consumer expectations about chicken soup products. In addition, Kano analysis was applied to identify how different features of chicken soup products influence consumer satisfactions about chicken soups.

2. Materials and Methods

2.1. Soup Preparation

The chicken soup stock was prepared weekly by boiling 1 kg drumsticks (separated into skins, flesh, and bones) in 3.5 L water in Crock-Pot with locking lid (Sunbeam Products, Inc., Boca Raton, FL, USA). Vegetables and spice (0.1 kg carrot, 0.1 kg yellow onion, 0.05 kg celery stalk, and 0.1 g bay leaf) were put on a four-layer cheese clothes in a tightly closed Crock-Pot. The stock was cooked at high heat for five hours and at low heat for three hours. Then the cheese clothes were immediately pulled from Crock-Pot and the stock was filtered through a four-layer cheese clothes again to generate a clear appearance. To balance the possible variations among the Crock-Pots, a completely randomized process was utilized to transfer the stock from Crock-Pot to glass jars: about 100–110 mL of stock was taken from individual Crock-Pot to fill up one 1.89 L glass jar. The jars were tightly sealed, cooled to room temperature, and stored at 4 °C for ~12 hours to remove top fat. The maximum storage of stock was four days at 4 °C. The chicken soup samples were prepared two hours before each sensory test. Table 1 presents the levels of NaCl and flavor enhancer in 1.5 L stock to make the soup. The usage level was selected based on the Equivalent Umami Concentration (EUC) of umami substances of each ingredient and their sensory enhancement effects on the chicken soup sample (data submitted elsewhere). The ingredients were added to 1.5 L of the stock at the beginning of the soup preparation and completely dissolved in the stock by string. Chicken soup was then heated at high heat in tightly closed Crock-Pot for one hour and 10 min and then kept at warm until the end of tests.

Table 1. Percentage (*w/v*) of salt (NaCl) and flavor enhancer in 1.5 L chicken soup.

Sample	Flavor Enhancer		Salt (%)
	Name	Level (%)	
Control	-	-	0.4%
MC 0.1%	Mushroom concentrate	0.1%	0.4%
MSG 0.1%	Monosodium glutamate	0.1%	0.4%
Salt 0.5%	-	-	0.5%
TC 0.2%	Tomato concentrate	0.2%	0.4%
YE 0.025%	Yeast extract	0.025%	0.4%

MC, mushroom concentrate; MSG, monosodium glutamate; TC, tomato concentrate; YE, yeast extract.

2.2. Degree of Difference from Control Testing

Approval from UGA's IRB (Project ID: STUDY00004396) was obtained before collecting the sensory data.

A degree of difference from control (DODC) test was carried out to find flavor differences in the test chicken soup samples in comparison to the control sample. A DODC test is a combination of discrimination and descriptive tests. Two samples are served together as in discrimination test (one control and one test sample) to examine if the samples are different. But it has another layer similar to descriptive analysis, where the source of differences is identified and the intensity of each difference is measured. References are not needed for a DODC test since control or reference sample is always served with the test sample. A total of six panelists who have more than 10 years of experience in descriptive analysis and one year experience in DODC tests were recruited for this study. Eight two-hour training sessions were conducted on the test samples to familiarize the panel with the major flavor attributes and to calibrate the panel's performance. Eight impact attributes were identified through panel consensus, which were overall flavor, meaty flavor, chicken flavor, salty taste, umami taste, yeast flavor, mushroom flavor, and tomato flavor. A pair of samples were served together: one was the test sample labeled with 3-digit code and the other was marked as "Control", which was the chicken soup with 0.4% salt. Panelists were required to compare the test sample against the "Control" for each attribute to evaluate the differences, if any. Each attribute was first rated on a comparison scale anchored at −1, 0 and 1, which represented "less than the Control", "same as the Control", and "greater than the Control" respectively. A 0 to 10-point intensity scale with 0.5 increments was then used to scale the degree of difference between the test sample and "Control". The values of 2.5, 5, and 7.5 were labeled as "slight difference", "moderate difference", and "large difference", respectively, on the intensity scale. About 60 mL of soup was served at 50 ± 2 °C in a 118-mL Styrofoam cups (Dart Container, Mason, MI, USA) wrapped with aluminum foil to avoid temperature loss during serving. The samples were tested in a random order and replicated twice. Six samples were evaluated against "Control" in each testing session. There was a two-minute break between two sets of samples to avoid panel fatigue. Unsalted Saltine crackers (Nabisco, East Hanover, NJ, USA) and deionized water were used as palate cleansers.

2.3. Consumer Testing of Chicken Soup

All the 6 samples were evaluated by 93 consumers who were recruited from an existing consumer database maintained at Sensory Evaluation and Consumer Lab, Department of Food Science and Technology, University of Georgia (Griffin Campus). All the consumers were recruited based on the following criteria: (1) 18 years of age and above, (2) 30% male, (3) no food allergies or intolerances, (4) shared at least equal responsibility of grocery shopping with other family members in the household, and (5) purchased and ate chicken soup products at least four times in the past year. Compusense® Cloud (Compusense, Inc., Guelph, ON, Canada) was used for data collection.

2.3.1. Taste Test

The consumer tests were carried out in partitioned booths under incandescent light at ~21 °C. About 60 mL of each chicken soup sample was served at ~50 ± 2 °C in a 118-mL Styrofoam cup. The samples were coded with three-digit random numbers and served in a sequential monadic order based on a completely randomized design. Unsalted Saltine crackers and distilled water were served as palate cleansers in-between samples. A nine-point hedonic scale (1—dislike extremely; 5—neither like nor dislike; 9—like extremely) was used for liking questions and a seven-point Just-about-right (JAR) scale (1—too weak; 4—JAR; 7—too strong) was used for color, overall flavor, salty, and savory (umami) tastes intensity questions.

2.3.2. Expectation Test on Ingredient List

Five lists of ingredients for chicken soup samples were presented to consumers in a fixed sequential monadic order from Panel A to E (Table 2). According to the information on the ingredient list, consumers were firstly required to rate their opinions of "free of artificial ingredients" for the chicken soup product on a "yes/no/maybe" categorical scale. If consumers selected "yes", they would rate their perceptions about "natural" on the same categorical scale; otherwise, they were required to indicate which ingredient(s) made them consider that the soup product was or might be "artificial". In addition, consumers' interest in consuming the chicken soup was evaluated on a five-point scale (1—definitely would not consume; 3—might or might not consume; 5—definitely would consume).

Table 2. Ingredient lists of chicken soup samples used in consumer testing.

Panel	Abbreviation	Ingredient Panel
A	MSG	Water, chicken, carrot, onion, celery, salt, monosodium glutamate, bay leaf
B	MC	Water, chicken, carrot, onion, celery, salt, mushroom concentrate, bay leaf
C	TC	Water, chicken, carrot, onion, celery, salt, tomato concentrate, bay leaf
D	YE	Water, chicken, carrot, onion, celery, salt, yeast extract, bay leaf
E	Salt	Water, chicken, carrot, onion, celery, salt, bay leaf

2.3.3. Kano Analysis

There were 14 features about chicken soup products that were tested (Appendix A). Majority of them were from claims of commercial products and from the results of our previous study [7]. Consumers were asked to answer three questions for each feature. The first question was a functional form, which captured the consumer's response to the presence of the attribute. The second question was a dysfunctional form that captured the consumer's response if that attribute was absent. A five-point scale was applied with labels in the order of "I will like it", "I must have it", "I don't care", "I can live with it" and "I will dislike it" from left to right. The third question was a nine-point self-stated importance scale (1—not at all important; 3—somewhat important; 5—important; 7—very important; 9—extremely important).

2.3.4. Demographic and Behavior Questionnaire

In the last part of the questionnaire, demographic data were collected. Consumer's concern level about MSG was evaluated on a seven-point scale [7]. In addition, some consumption and purchase questions about chicken soup products were asked.

2.4. Statistical analysis

The data from DODC and consumer tasting test were analyzed by two-way and one-way analysis of variance (ANOVA) respectively in SAS (version 9.4, SAS Institute, Cary, NC, USA) using the GLIMMIX procedure (General Linear Mixed Models). Panelist was considered as a random factor. Least square means were calculated. Post-hoc mean separation was done using Fisher's least significant difference ($P < 0.05$). JAR data were grouped to three categories that were "too little" (<4), JAR (4), and "too much" (>4). Their percentage distributions were reported. One-sample t-test was then conducted in SAS to compare individual product's mean JAR intensity versus the ideal JAR intensity which was 4 on the scale.

For Kano analysis, this study used the continuous analysis method proposed by Bill DuMouchel, which solved the challenges in traditional discrete analysis method like misclassification caused by similar counts for several categories and incapability to distinguish the features under the same category [17]. The options of Kano scale were translated to a numeric value as follows:

Functional (Y): −2 (Dislike), −1 (Live with), 0 (Don't care), 2 (Must-have), 4 (Like);

Dysfunctional (X): −2 (Like), −1 (Must-have), 0 (Don't care), 2 (Live with), 4 (Dislike).

For each question, the averages of X (dysfunctional), Y (functional), and importance score were computed across the respondents. Only the averages that fall in the range of 0 to 4 were displayed on a scatter plot. The plot was divided into quadrants corresponding to one-dimensional, must-have, attractive, and indifferent (Figure 1). Then, the average of the importance score for each question was visualized by converting its scatter plot dot to bubble with size proportional to its importance.

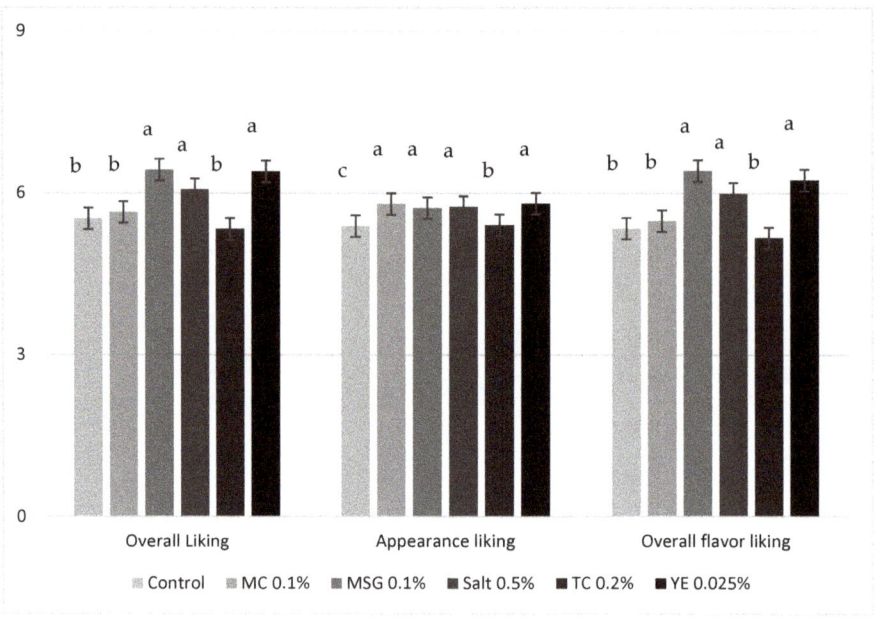

Figure 1. Mean consumer liking scores for chicken soups. MC, mushroom concentrate; MSG, monosodium glutamate; TC, tomato concentrate; YE, yeast extract.

3. Results and Discussion

3.1. Degree of Difference from Control (DODC) Testing

Eight attributes were measured in this test and significant differences ($P < 0.05$) were detected for seven of them (Table 3). Only yeast flavor, the identity (ID) flavor of YE, was statistically same ($P \geq 0.05$) among six chicken soup samples. The mean degree-of-difference-from-control (DODC) intensities of the eight attributes for all the samples are presented in Table 3. The "Control" sample was compared against the "Control" reference and its DODC intensities were "zero" for all the attributes, which implied the accuracy of the descriptive panel. Addition of individual flavor enhancer (MC 0.1%, MSG 0.1%, TC 0.2%, YE 0.025%) significantly enhanced the overall flavor, meaty flavor, saltiness, and umami taste of the chicken soup added with 0.4% salt. YE 0.025% showed statistically same enhancement effect as MSG 0.1% in all the attributes. MC 0.1% sample exhibited a significantly weaker improvement in the chicken flavor and umami taste than MSG 0.1% but had a stronger mushroom ID flavor. This meant that higher concentrations of MC would contribute more mushroom flavor to the soup instead of enhancing the chicken flavor. This phenomenon was also observed with TC. TC was used at the highest level among flavor enhancers which successfully boosted the overall flavor, meaty flavor, salty taste, and umami taste of chicken soup to the level achieved by MSG 0.1%. However, it also significantly increased the tomato ID flavor. In addition, TC 0.2% sample had a significantly lower chicken flavor than MSG 0.1% and YE 0.025% and lower salty taste than YE 0.025%. In addition to glutamate, YE's higher nucleotide content might be the reason for its high umami perceptions and enhancement effects (data submitted elsewhere). Jo and Lee came to similar conclusions as ours while comparing commercial yeast extracts and their combinations with MSG in water solutions [9].

Salt is crucial to the palatability of a wide variety of foods. Besides increasing the saltiness, salt also contributes to the overall flavor and suppression of bitter taste [18]. In this study, Salt 0.5% sample had a significantly higher ($P < 0.05$) DODC value of most attributes than Control (0.4% salt sample), suggesting that usage of extra 0.1% of salt improved the flavors and tastes of chicken soup. On the other hand, 20% salt reduction was already able to significantly impact the sensory profile of chicken soup. It is well known that umami taste enhances saltiness and other flavor attributes. Previous research studies have reported that addition of yeast extract could reduce sodium usage without sacrificing sensory pleasantness [12,19]. Our results indicated that YE 0.025% and MSG 0.1% had similar effects as Salt 0.5% on overall flavor and salty taste. Compared to Salt 0.5% sample, our panelists perceived significantly stronger meaty flavor, chicken flavor, and umami taste in YE sample. Although MC 0.1% and TC 0.2% samples had significantly lower ($P < 0.05$) salty taste than Salt 0.5% sample, they still showed a similar or even higher intensity in majority of the attributes. Overall, our findings proved that all the four flavor enhancers at current level could compensate for the flavor and taste loss caused by 20% salt reduction.

Table 3. The degree-of-difference-from-control (DODC) mean intensity scores (standard deviation in parentheses) of chicken soup samples on a 10-point intensity scale.

Soup	Overall Flavor	Meaty Flavor	Chicken Flavor	Salty Taste	Umami Taste	Yeast Flavor	Mushroom Flavor	Tomato Flavor
Control [1]	0 (0) [b,2]	0 (0) [c]	0 (0) [c]	0 (0) [c]	0 (0) [d]	0 (0)	0 (0) [b]	0 (0) [c]
MC 0.1% [1]	3.1 (0.4) [a]	3.0 (0.4) [a]	0.8 (1.3) [bc]	2.5 (0.6) [b]	2.6 (0.6) [bc]	0.1 (0.3)	1.2 (1.3) [a]	0.1 (0.3) [c]
MSG 0.1% [1]	3.2 (0.4) [a]	2.6 (0.7) [ab]	2.2 (0.7) [a]	2.9 (0.4) [ab]	3.1 (0.4) [a]	0.3 (0.7)	0.4 (0.7) [b]	0.5 (0.7) [b]
Salt 0.5%	3.3 (0.2) [a]	2.2 (1.1) [b]	0.8 (1.1) [bc]	3.3 (0.3) [a]	2.4 (0.9) [c]	0.2 (0.5)	0.4 (0.6) [b]	0.1 (0.3) [c]
TC 0.2% [1]	3.1 (0.4) [a]	2.6 (0.7) [ab]	1.2 (1.4) [b]	2.5 (0.8) [b]	2.8 (0.3) [ab]	0 (0)	0.3 (0.5) [b]	2.5 (0.5) [a]
YE 0.025% [1]	3.3 (0.4) [a]	3.0 (0.4) [a]	2.1 (0.8) [a]	3.1 (0.5) [a]	3.0 (0.4) [ab]	0.6 (1.4)	0.1 (0.3) [b]	0.3 (0.6) [bc]

[1] Contained 0.4% salt; [2] Different letters in the same column indicates significant difference ($P < 0.05$).

3.2. Consumer Taste Test

A total of 93 consumers participated in the test. The mean intensity of three hedonic responses were significantly different among six chicken soup samples (Figure 1). Same liking pattern was observed for overall liking and overall flavor liking. MSG 0.1% had highest score for these two hedonic attributes, followed by YE 0.025%. In general, consumers slightly liked MSG 0.1%, YE 0.025%, and Salt 0.5% sample with no statistical difference among them. They also obtained significantly higher ($P < 0.05$) acceptability scores than Control, MC 0.1%, and TC 0.2%. The capability of MSG in sensory enhancement has been widely proved in different food product. Ribonucleotides (mainly IMP and GMP) also contribute to palatability elevation, which exhibit synergistic influences with MSG. Miyaki et al. reported that the overall liking of chicken noodle soup was significantly increased ($P < 0.05$) by using 0.1% MSG or 0.3% MSG + 0.1% IMP, and the sample enhanced by IMP and MSG together was liked most among all the samples [6]. Baryłko-Pikielna and Kostyra studied the effects of MSG (0–0.5%) and IMP +GMP (0–0.015%) on the palatability of seven model food matrices [4]. Application of MSG and/or IMP+GMP increased the hedonic response in all model products, but the degree of elevation varied considerably among the products. In their work, MSG played a leading role in the increased acceptability of chicken soup. Yeast extract is rich in both glutamate and ribonucleotides. Therefore, it was not surprising to find that compared to MSG, YE increased the palatability of chicken soup to a similar level at a much lower dose. Mushrooms and tomatoes are also very rich in umami substances. However, their extracts failed to significantly increase the overall liking and overall flavor liking of the chicken soup. When cross-compared to the DODC findings, although MC 0.1% and TC 0.2% sample had significantly higher values for all the sensory attributes than Control, they still showed a significantly lower chicken flavor than MSG 0.1% and YE 0.025%, and lower saltiness than Salt 0.5% and YE 0.025%. Therefore, a relatively lower chicken flavor and salty taste might have caused their failure in acceptability improvement. Previous studies have also reported a positive relationship between chicken flavor and overall liking of chicken meat and stock products [20,21]. In addition, TC 0.2% had a DODC intensity of 2.5 for tomato flavor which indicated a slight difference from Control. This slightly higher tomato flavor could also adversely affect consumer acceptability of chicken soup.

Top two boxes score (combined percentage for the responses of "like extremely" and "like very much") for overall liking represents the percentage of consumers that strongly liked the products. Its findings were generally in an agreement with liking means (data not presented). More than 20% of participants generally liked Salt 0.5% (25.8%), MSG 0.1% (23.7%), and YE 0.025% sample (23.7%). The top 2 box percentage for Control and MC 0.1% was 14.0% and 16.1% respectively. Only 8.6% of consumers liked TC 0.2% extremely or very much. In contrast to the Control, all the enhanced chicken soup samples had a significantly higher appearance liking. Consumer did not perceive any difference in the appearance liking among MC 0.1%, MSG 0.1%, Salt 0.5%, and YE 0.025%. All these four samples were liked significantly more than TC 0.2% appearance-wise. The difference in appearance likings might origin from the Halo effect of overall liking. Overall, palatability of chicken soup could be improved by adding 0.1% MSG, 0.025% YE, or extra 0.1% salt. This increased acceptability might result from increased chicken flavor and/or salty taste.

Consumers also evaluated the optimal intensity of color, overall flavor, salty, and savory (umami) tastes using seven-point JAR scale (Figure 2a–d). Samples that were liked more by consumers also got a larger percentage for JAR category (4 on the scale). Similar distribution pattern of JAR data was observed among overall flavor, salty, and savory tastes of six chicken soups. YE 0.025%, MSG 0.1%, and Salt 0.5% consistently had lesser "too low" and more JAR scores than other samples for these three attributes. The pattern for color JAR was slightly different because the scores for MC 0.1% were almost evenly distributed among three categories. Eighty percent of JAR scores is considered as a common benchmark by some product developers [22]. In this study, most of the JAR scores were less than 50%, which was far below the benchmark of 80%. One-sample t-test was also conducted to compare individual product's mean JAR intensity versus the center point's value on the scale (4) for each attribute. A significantly lower value was detected for all the tests, indicating the mean intensity was very much skewed below the appropriate level.

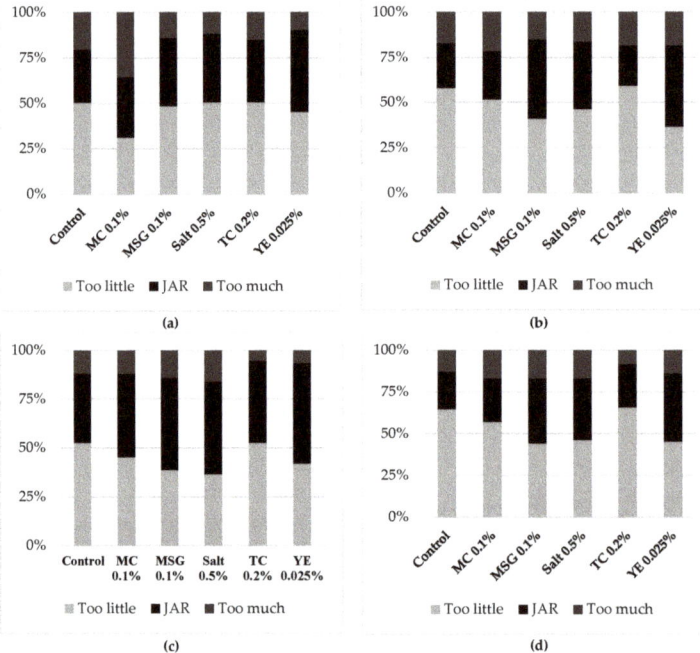

Figure 2. Distribution of consumer JAR scores for sensory attributes of chicken soups: (**a**) Color; (**b**) Overall flavor; (**c**) Salty taste; (**d**) Savory taste.

Despite all the samples not being close to the ideal, the enhancement effect was still noticeable. Control sample had smallest percentage of JAR for color, salty and savory tastes. Using either flavor enhancers or extra salt successfully increased the JAR scores of these three attributes but to different extents. YE 0.025% showed the largest improvement especially in the overall flavor and savory taste. Approximate two-fold increment in JAR percentage was achieved by addition of YE at 0.025% in contrast to Control. Salt 0.5% and MSG 0.1% had nearly same proportion of JAR for color, salty, and savory tastes respectively. TC 0.2% and MC 0.1% showed an extremely weak influence on the JAR of savory taste when compared to Control. As for JAR of overall flavor, using TC at 0.2% slightly reduced the percentage of appropriateness compared to Control. In DODC testing, TC 0.2% sample had stronger overall flavor than Control and no difference in this attribute from other enhanced samples. However, lowest proportion of consumers voted JAR for overall flavor of TC 0.2%, which was about half of that for YE 0.025%. The inconsistent findings of overall flavor in consumer and DODC study possibly result from different understandings of overall flavor among consumers and trained panelists.

3.3. Effect of Information on Consumer Perceptions

Food packaging and labelling are normally the first contact between consumer and a processed food product [14]. Since consumer's ingredient awareness for natural and healthy foods rise, food label plays a significant role in consumer food choices. As an important component of food label, an ingredient panel provides the information that might affect consumer perception about "naturalness". In this study, ingredient lists for the chicken soup samples used in the taste test were investigated among 93 consumers. Since Control and Salt 0.5% had the same ingredient information, only five lists were applied. Consumers firstly evaluated their perceptions about "free of artificial ingredients" for individual ingredient panel. If they considered the panel had or might have artificial ingredients, they were required to select all the ingredients that gave them the artificial perception. Table 4 shows that only 31.2% of participants perceived that Panel A (MSG) did not contain artificial ingredient, which was lowest. Another 47.3% considered MSG-contained chicken soup had artificial ingredients and MSG was the only ingredient selected by them for artificial perception. The rest of 21.5% of consumers thought Panel A might have artificial ingredients and 85% of them voted for MSG (data not presented). By changing MSG to its substitutes, more consumers considered the product as free of artificial ingredients. The degree of elevation followed the ascending order of Panel D (YE) < Panel B (MC) < Panel C (TC). The exact same order was observed in our previous study, which surveyed consumer's choice of MSG substitutes in chicken soup [7]. There were around 10–27% of consumers still considered the chicken soup with MSG substitutes were or might be artificial, and majority of them pointed to the corresponding alternative in the ingredient panel. Panel E (Salt) which did not include any flavor enhancer obtained highest percentage (91.4%) for "free of artificial". Previous studies found that consumers perceived less naturalness from food additive with hard-to-pronounce name, chemical name, or from less familiar source. Thus, the chemical name of MSG resulted in the highest responses for "artificial". In addition, consumers might be more familiar with tomato and mushroom in food, leading to a higher percentage of "artificial-free" than YE. Rozin indicated that using additives is associated with "chemical transformation", which destroyed the naturalness more compared to physical changes like grinding and freezing [23]. This likely explains why more participants considered the enhancer-free ingredient panel as free of artificial ingredients.

The perception of naturalness was further explored for the consumers who considered the product did not have any artificial ingredient. Similar to "free of artificial ingredient" perception, the chicken soup products considered least and most natural was Panel A (MSG) and Panel E (salt) correspondingly (Table 4). According to our previous study, the reasons for artificial image of MSG included "not present naturally in food", "its name and/or its abbreviation", and "manufactured from inedible materials/chemicals" [7]. This, on the other hand, explained why chicken soup contained MSG substitute had higher natural perception. Other scientists also reported that only 2% respondents considered MSG natural and concluded that consumers did not know what ingredient was and was not natural with various reasons to justify their natural perception, which introduced significant difficulties for FDA to come up with an acceptable definition agreed by consumer, academia, and producer [24]. As shown in Table 4, the percentage for "natural" was consistently less than that for "free of artificial" for individual ingredient panel. This implies that "free of artificial ingredients" might only be a component of consumer "natural perception".

Table 4. Consumer perception about the five ingredient panels for chicken soup products.

Panel [1]	Abbreviation	Free of Artificial			Natural [2]			Interest Intensity for Consumption [3]
		Yes	No	Maybe	Yes	No	Maybe	
A	MSG	31.2%	47.3%	21.5%	18.3%	9.7%	3.2%	3.7 c,[4]
B	MC	65.6%	20.4%	14.0%	50.6%	7.5%	7.5%	4 b
C	TC	76.3%	9.7%	14.0%	62.3%	6.5%	7.5%	3.9 b
D	YE	50.5%	22.6%	26.9%	36.5%	7.5%	6.5%	3.5 d
E	Salt	91.4%	6.5%	2.2%	87.0%	2.2%	2.2%	4.5 a

[1] Refer to Table 3 for the detailed information about the ingredient panel; [2] Only consumer who selected "Yes" for "Free of artificial ingredient" answered this question; [3] Consumer rated their level of interest to consumer on a 5-point Likert-like intensity scale; [4] Different letters in the same column indicates significant difference ($P < 0.05$).

Consumers also rated their interest in consuming the chicken soup for corresponding ingredient panel. Consumers had the strongest interest in the flavor enhancer–free product, which was significantly higher ($P < 0.05$) than the products containing enhancer (Table 4). Among the enhancers, consumers possessed a significantly higher ($P < 0.05$) interest in TC and MC-enhanced product. YE containing product had a value, which was significantly lower than the rest. A mismatch was noticed when comparing the results of taste test to label perception test. YE- and MSG-added chicken soup, which consumers liked most in taste test obtained the lowest interest in label perception test. This discrepancy also existed in TC and MC sample. When consumers did not taste the chicken soup, their interest in trying the product seemed to correlate to their perceptions of naturalness and free-of-artificial-ingredient about the corresponding ingredient panel. The decision about everyday food choice is a trade-off between various sensory and non-sensory factors [13]. The effects of information on consumer perceptions about food products have been investigated by many researchers. Bandara et al. reported that list of ingredients was ranked as the second most important mandatory labeling information by consumers and only 3% respondents' purchase intent was not influenced by MSG content [15]. Besides ingredient, family influence, price, and taste also drive consumer attitudes towards food products [25]. It is likely that consumers would imagine the taste of products by looking at the information on the package [14]. We speculate that the unpleasant taste imagined by consumers for yeast extract might result in the lowest interest in consuming the corresponding chicken soup product. Prescott and Young found that the information about MSG usage did not influence consumer liking, natural taste perception, and purchase intent of vegetable soup when the product and label information were served together to consumers [16]. They claimed that when tasting the products, consumers weighted sensory properties more important than information highly relevant to their beliefs and attitudes. Due to the financial and time limitation, we did not test the effect of ingredient information on hedonic responses. However, based on previous study, information about taste might increase consumer willingness to try food products [26].

Considering that the habit of reading ingredient panel during shopping for chicken soup products might affect consumer perceptions about the ingredient information, consumers were further separated to two groups. The first group contained 56 consumers who read or sometimes read ingredient panel of chicken soup products during shopping; the other 37 consumers did not have this shopping habit. The percentages of selection for "free of artificial ingredients" and "natural" were calculated for each ingredient panel, no difference was noted among two groups (data not presented). ANOVA was also conducted to analyze the differences in the interest of consumption between the two groups. No significant group effect or group by ingredient effect was detected (data not presented).

These findings indicated that whether consumer read ingredient list or not they had no difference in free-of-artificial and natural perception and consumption interest in in chicken soup products.

3.4. Kano Analysis

Kano questions are used to understand the attributes affecting consumer satisfaction or dissatisfaction about a product. There are six categories for product features in the Kano model [17]. Must-have feature is the attribute expected by customers and sometimes taken for granted when fulfilled. Its absence or poorly satisfaction results in large customer dissatisfaction. One-dimensional feature has a positive linear relationship to customer's satisfaction. Attractive feature provides great satisfaction when fulfilled but are acceptable when not fulfilled. Indifferent attribute does not affect consumer satisfaction. Reverse feature results in dissatisfaction when fulfilled and in satisfaction when not fulfilled. When a consumer has conflicting responses, the attribute goes to the category of questionable feature.

In this study, none of the 14 attributes belonged to questionable or reserve category. "Taste good" was the only one-dimensional attribute for chicken soup product (Figure 3). Tasting good is a must-have attribute for bacon but a one-dimensional attribute for cottage cheese [27,28]. "Free of artificial ingredient", "free of added MSG", "all natural", "free of preservative", "ready to serve", and "lower price" were attractive attributes for chicken soup products. Most of these attractive attributes were relevant to food additives. The rest of the seven attributes were all in the group of indifferent attributes. Similar as our study, consumer also indicated "organic", "low sodium" and "fat-free" were indifferent features for cottage cheese [28]. As an important category of chicken soup products, it was surprising to find that "low sodium" was indifferent to consumers. It seems that this group of consumers cared more about artificial or natural ingredient usage than sodium reduction in chicken soups. Both ready-to-serve and on-the-go packaging are convenience-related features. Consumers considered "ready-to-serve" attractive but "on-the-go" packaging indifferent. Normally, feature classification transitions in the direction of indifferent to attractive to one-dimensional and finally to must-have due to market maturity [29]. Since "on-the-go" is a relatively new concept in chicken soup products, it is likely that consumers might value it more as its growth in the market.

The self-stated importance rating was firstly compared among all attributes using ANOVA and then projected on to the Kano plot with the size of individual attributes proportional to its importance. "Taste good" was the most important attribute to consumers with a score of 8.1 on a 9-point scale, which was significantly higher ($P < 0.05$) than all the other attributes (data not presented). "Free of added MSG" was the second important attribute of chicken soup products. Consumer perceived no difference in the importance of "free of added MSG", "all natural", "lower price", "GMO free", "free of artificial ingredients" and "free of preservatives." But only "free of added MSG" obtained a score of 5 which was the benchmark for important attributes on a 9-point scale. Therefore, "Free of added MSG" was the only attractive feature that consumers felt important. According to FDA, foods with any ingredient that naturally contains MSG cannot claim "No MSG" or "No added MSG" on their packaging [30]. However, there are some manufacturers still claiming "No (added) MSG" on either packaging, website, or other sources for soup products enhanced by MSG substitutes that naturally contain MSG. More clear definition and stronger regulation may be required to clarify MSG-related claims. According to Figure 3, the self-stated importance generally decreased in the order of one-dimensional > attractive > indifferent. However, GMO-free, an indifferent feature, obtained an importance score that was statistically same to most of attractive attributes. "Ready-to-serve" which was attractive to consumers grouped with other indifferent attributes from viewpoint of self-stated importance.

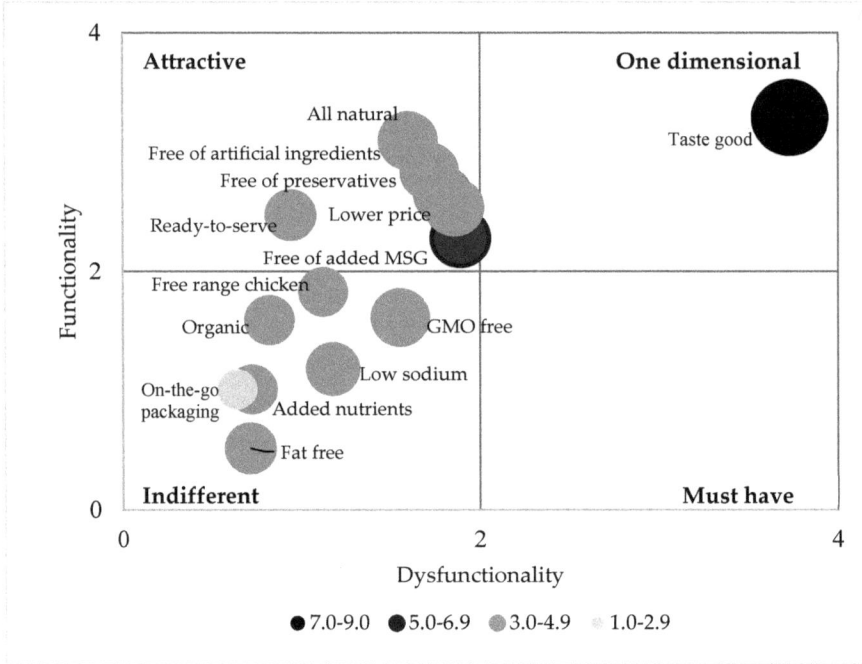

Figure 3. Kano classifications for 14 features about chicken soup products with averaged importance for each attribute indicated by size and color of the bubble.

3.5. Influence of MSG Concern Level on Responses

Consumers also indicated their concern level about MSG usage in food products on a seven-point Likert-like scale. Based on their responses, consumers were clustered into three groups: the first group was the concerned group which consisted of 43 consumers who rated their concern level at 5 and above; the second group was the unconcerned group which had 33 consumers whose concern level was below 5; and the rest of 17 consumers did not hear about MSG before, so they were considered as a separate group to avoid bias. The average concern level for the first group was 5.7, which was significantly higher than that of the second group (3.2), and third group (4.1). Consumers in the third group were also concerned more than consumers in the second group. These three groups were then compared for any difference in the other responses. No difference among three groups was detected for overall liking of the six chicken soup samples and perceptions about the individual ingredient panel. As for Kano analyses, most of findings aligned with previous results. "Free of preservatives" became one-dimensional feature to consumers in concern group. Consumers who did not hear about MSG perceived "added nutrients", "free range chicken" and "on-the-go packaging" significantly more important than consumer in concern group did.

3.6. Soup Eating and Purchasing Behavior Questions

Behavior questions about chicken soup products were also investigated in this work. Consumers were firstly required to select top three health benefits that they would like to see most in chicken soup products. Higher protein, added vegetables, and low sodium were the top three choices with 69.9%, 62.4%, and 49.5% respectively. Other benefits included low calorie (41.9%), no trans-fat (38.7%), added vitamins (24.7%), and added minerals (12.9%). It seems that using more vegetables in chicken soup could be a good strategy to increase consumer intake of vitamins and minerals. Consumers who selected low sodium in previous question were further asked about the level of sodium reduction they wanted most in chicken soup. Results showed that 43.5% liked 30–50% sodium reduction and 26.1% voted for less than 30% sodium reduction. Thus, majority of participants preferred 50% and less sodium reduction in chicken soup products. As for the reasons or motivations for chicken soup consumption, 68.8% expressed that they ate chicken soups when they are sick. Around 63.4% considered chicken soup easy and fast to prepare; another popular reason was the affordability (44.1%) of chicken soups. Other motivations and reasons were "personal preference" (38.7%), "maintain a healthy diet" (36.6%), and "family tradition" (24.7%). Ready-to-eat was the type of chicken soup products that most consumers (80.6%) usually purchased. Condensed (67.7%) and wet broth/stock (65.6%) also had more than 60% selection. Dry and frozen/refrigerated chicken soup products were less popular among this group of consumers with a percentage of 29% and 10.8% respectively.

4. Conclusions

Overall, MSG and all of its alternatives enhanced the flavors and tastes of chicken soup with 0.4% salt. YE 0.025% had the same effects as MSG 0.1% in all tested sensory attributes. All the flavor-enhanced samples had the same or even stronger overall flavor, meaty flavor, chicken flavor, and umami taste when compared to chicken soup added with 0.5% salt. But addition of MC 0.1% and TC 0.2% failed to increase the salty taste to the level achieved by using 0.5% salt. Consumers significantly preferred MSG 0.1%, YE 0.025%, and Salt 0.5% samples to the others. Information about ingredient usage largely affected consumer perceptions about the chicken soup. Our results showed that familiarity bias toward ingredient name played a significant role in consumer "free of artificial" and "natural" perception and interest to consume chicken soup products. MSG and yeast enhanced sample which had higher consumer likings in taste test obtained a lower consumer interest in ingredient panel testing. Kano analysis indicated that "good taste" was the most important feature influencing consumer satisfaction about chicken soups. Claims related with artificial and natural ingredient were attractive attributes to consumers. Our findings suggested that using natural extracts, especially yeast extract, could successfully compensate for the palatability loss caused by removal of MSG or 20% salt reduction. However, there was a mismatch between consumer sensory liking and their expectations based on ingredient usage. More studies are required to investigate how information about ingredient usage changes consumer hedonic responses towards chicken soup products.

Author Contributions: S.W. designed the study, collected the data, interpreted the results, and wrote the manuscript. S.Z. contributed to Kano analysis and its interpretation. K.A. provided guidance for the study, helped with the interpretation of the data, and edited the manuscript.

Funding: This research received no external funding.

Acknowledgments: Thanks are due to the Graduate School, the University of Georgia, for supporting this study.

Conflicts of Interest: The authors declare no conflict of interest.

Appendix A. Kano Analysis Questionnaire

For each feature, the first two questions are on a 5-point scale where 1 = I will like it; 2 = I must have it; 3 = I don't care; 4 = I can live with it; 5 = I will dislike it. The third question was on a 9-point scale where 1 = Not at all important; 3 = somewhat important; 5 = Important; 7 = Very important; 9 = extremely important.

1. Artificial Ingredients Free

If the chicken soup product is free of artificial ingredients, how do you feel?
If the chicken soup product is not free of artificial ingredients, how do you feel?
How important it is or would it be if the chicken soup product is free of artificial ingredients?

2. *Added Monosodium Glutamate (Msg) Free*

>   If the chicken soup product is free of added monosodium glutamate (MSG), how do you feel?
>   If the chicken soup product is not free of added monosodium glutamate (MSG), how do you feel?
>   How important it is or would it be if the chicken soup product is free of added MSG?

3. *All Natural*

>   If the chicken soup product is all-natural, how do you feel?
>   If the chicken soup product is not all-natural, how do you feel?
>   How important it is or would it be if the chicken soup product is all-natural?

4. *Low Sodium*

>   If the chicken soup product is low in sodium, how do you feel?
>   If the chicken soup product is not low in sodium, how do you feel?
>   How important it is or would it be if the chicken soup product is low in sodium?

5. *Preservative Free*

>   If the chicken soup product is free of preservatives, how do you feel?
>   If the chicken soup product is not free of preservatives, how do you feel?
>   How important it is or would it be if the chicken soup product is free of preservatives?

6. *Taste Good*

>   If the chicken soup product tastes good, how do you feel?
>   If the chicken soup product does not taste good, how do you feel?
>   How important it is or would it be if the chicken soup product tastes good?

7. *Fat-Free*

>   If the chicken soup product is fat-free, how do you feel?
>   If the chicken soup product is not fat-free, how do you feel?
>   How important it is or would it be if the chicken soup product is fat-free?

8. *Added Nutrients*

>   If the chicken soup product has added nutrients, how do you feel?
>   If the chicken soup product does not have added nutrients, how do you feel?
>   How important it is or would it be if the chicken soup product has added nutrients?

9. *Genetically Modified Organism (Gmo) Free*

>   If the chicken soup product is GMO-free, how do you feel?
>   If the chicken soup product is not GMO-free, how do you feel?
>   How important it is or would it be if the chicken soup product is GMO-free?

10. *Organic*

>   If the chicken soup product is organic, how do you feel?
>   If the chicken soup product is not organic, how do you feel?
>   How important it is or would it be if the chicken soup product is organic?

11. *Ready to Serve*

>   If the chicken soup product is ready to serve, how do you feel?
>   If the chicken soup product is not ready to serve, how do you feel?
>   How important it is or would it be if the chicken soup product is ready to serve?

12. Free range Chicken

If the chicken soup product uses free range chicken, how do you feel?
If the chicken soup product does not use free range chicken, how do you feel?
How important it is or would it be if the chicken soup product uses free range chicken?

13. On-the-Go Packaging

If the chicken soup product is in on-the-go packaging, how do you feel?
If the chicken soup product is not in on-the-go packaging, how do you feel?
How important it is or would it be if the chicken soup product is in on-the-go packaging?

14. Price

If the chicken soup product has price lower than other chicken soup products in the store you always shop, how do you feel?

If the chicken soup product does not have price lower than other chicken soup products in the store you always shop, how do you feel?

How important it is or would it be if the chicken soup product has price lower than other chicken soup products in the store you always shop?

References

1. Grand View Research. Available online: https://www.grandviewresearch.com/press-release/us-soup-market-analysis (accessed on 8 October 2018).
2. Saketkhoo, K.; Januszkiewicz, A.; Sackner, M.A. Effects of drinking hot water, cold water, and chicken soup on nasal mucus velocity and nasal airflow resistance. *Chest* **1978**, *74*, 408–410. [CrossRef]
3. Rennard, B.O.; Ertl, R.F.; Gossman, G.L.; Robbins, R.A.; Rennard, S.I. Chicken soup inhibits neutrophil chemotaxis in vitro. *Chest* **2000**, *118*, 1150–1157. [CrossRef] [PubMed]
4. Baryłko-Pikielna, N.; Kostyra, E. Sensory interaction of umami substances with model food matrices and its hedonic effect. *Food Qual. Prefer.* **2007**, *18*, 751–758. [CrossRef]
5. Dermiki, M.; Prescott, J.; Sargent, L.J.; Willway, J.; Gosney, M.A.; Methven, L. Novel flavours paired with glutamate condition increased intake in older adults in the absence of changes in liking. *Appetite* **2015**, *90*, 108–113. [CrossRef] [PubMed]
6. Miyaki, T.; Retiveau-Krogmann, A.; Byrnes, E.; Takehana, S. Umami increases consumer acceptability, and perception of sensory and emotional benefits without compromising health benefit perception. *J. Food Sci.* **2016**, *81*, S483–S493. [CrossRef] [PubMed]
7. Wang, S.; Adhikari, K. Consumer perceptions and other influencing factors about monosodium glutamate in the United States. *J. Sens. Stud.* **2018**, *33*, e12437. [CrossRef]
8. Sombutyanuchit, P.; Suphantharika, M.; Verduyn, C. Preparation of 5'-GMP-rich yeast extracts from spent brewer's yeast. *World J. Microbiol. Biotechnol.* **2001**, *17*, 163–168. [CrossRef]
9. Jo, M.N.; Lee, Y.M. Analyzing the sensory characteristics and taste-sensor ions of MSG substitutes. *J. Food Sci.* **2008**, *73*, S191–S198. [CrossRef]
10. Myrdal Miller, A.; Mills, K.; Wong, T.; Drescher, G.; Lee, S.M.; Sirimuangmoon, C.; Schaefer, S.; Langstaff, S.; Minor, B.; Guinard, J.X. Flavor-enhancing properties of mushrooms in meat-based dishes in which sodium has been reduced and meat has been partially substituted with mushrooms. *J. Food Sci.* **2014**, *79*, S1795–S1804. [CrossRef]
11. Umami Information Center. Available online: http://www.umamiinfo.com/ (accessed on 8 October 2018).
12. Campagnol, P.C.B.; dos Santos, B.A.; Wagner, R.; Terra, N.N.; Pollonio, M.A.R. The effect of yeast extract addition on quality of fermented sausages at low NaCl content. *Meat Sci.* **2011**, *87*, 290–298. [CrossRef]
13. Jaeger, S.R. Non-sensory factors in sensory science research. *Food Qual. Prefer.* **2006**, *17*, 132–144. [CrossRef]
14. Carrillo, E.; Varela, P.; Fiszman, S. Packaging information as a modulator of consumers' perception of enriched and reduced-calorie biscuits in tasting and non-tasting tests. *Food Qual. Prefer.* **2012**, *25*, 105–115. [CrossRef]

15. Bandara, B.E.S.; De Silva, D.A.M.; Maduwanthi, B.C.H.; Warunasinghe, W.A.A.I. Impact of food labeling information on consumer purchasing decision: with special reference to faculty of agricultural sciences. *Procedia Food Sci.* **2016**, *6*, 309–313. [CrossRef]
16. Prescott, J.; Young, A. Does information about MSG (monosodium glutamate) content influence consumer ratings of soups with and without added MSG? *Appetite* **2002**, *39*, 25–33. [CrossRef] [PubMed]
17. Berger, C.; Blauth, R.; Boger, D.; Bolster, C.; Burchill, G.; DuMouchel, W.; Pouliot, F.; Richter, R.; Rubinoff, A.; Shen, D.; et al. Kano's methods for understanding customer-defined quality. *Cent. Qual. Manag. J.* **1993**, *2*, 2–36.
18. Breslin, P.A.S.; Beauchamp, G.K. Salt enhances flavour by suppressing bitterness. *Nature* **1997**, *387*, 563. [CrossRef] [PubMed]
19. Mitchell, M.; Brunton, N.; Wilkinson, M. Optimization of the sensory acceptability of a reduced salt model ready meal. *J. Sens. Stud.* **2009**, *24*, 133–147. [CrossRef]
20. Lee, Y.S.; Owens, C.M.; Meullenet, J.F. On the quality of commercial boneless skinless broiler breast meat. *J. Food Sci.* **2008**, *73*, S253–S261. [CrossRef]
21. Kim, H.; Lee, J.; Kim, B. Development of an initial lexicon for and impact of forms (cube, liquid, powder) on chicken stock and comparison to consumer acceptance. *J. Sens. Stud.* **2017**, *32*, e12251. [CrossRef]
22. Rothman, L.; Parker, M.J. *Just-about-Right (JAR) Scales: Design, Usage, Benefits, and Risks*; ASTM International: West Conshohocken, PA, USA, 2009.
23. Rozin, P. The meaning of "natural": Process more important than content. *Psychol. Sci.* **2005**, *16*, 652–658. [CrossRef]
24. Chambers, E.V.; Tran, T.; Chambers, E., VI. "Natural": A consumer perspective. In Proceedings of the 2018 Society of Sensory Professionals Conference, Cleveland, OH, USA, 26–28 September 2018.
25. Kempen, E.; Bosman, M.; Bouwer, C.; Klein, R.; van der Merwe, D. An exploration of the influence of food labels on South African consumers' purchasing behavior. *Int. J. Consum. Stud.* **2011**, *35*, 69–78. [CrossRef]
26. Pelchat, M.L.; Pliner, P. "Try it. You'll like it": Effects of information on willingness to try novel foods. *Appetite* **1995**, *24*, 153–165. [CrossRef]
27. McLean, K.G.; Hanson, D.J.; Jervis, S.M.; Drake, M.A. Consumer perception of retail pork bacon attributes using adaptive choice-based conjoint analysis and maximum differential scaling. *J. Food Sci.* **2017**, *82*, 2659–2668. [CrossRef]
28. Hubbard, E.M.; Jervis, S.M.; Drake, M.A. The effect of extrinsic attributes on liking of cottage cheese. *J. Dairy Sci.* **2016**, *99*, 183–193. [CrossRef] [PubMed]
29. Kametani, T.; Nishina, K.; Suzuki, K. Attractive quality and must-be quality from the viewpoint of environmental lifestyle in Japan. In *Frontiers in Statistical Quality Control 9*; Lenz, H.-J., Wilrich, P.-T., Schmid, W., Eds.; Springer: Heidelberg, Germany, 2010; pp. 315–327.
30. U.S. Food & Drug Administration. Questions and Answers on Monosodium Glutamate (MSG). Available online: https://www.fda.gov/food/ingredientspackaginglabeling/foodadditivesingredients/ucm328728.htm (accessed on 8 October 2018).

© 2019 by the authors. Licensee MDPI, Basel, Switzerland. This article is an open access article distributed under the terms and conditions of the Creative Commons Attribution (CC BY) license (http://creativecommons.org/licenses/by/4.0/).

Article

The Effect of Emulsion Intensity on Selected Sensory and Instrumental Texture Properties of Full-Fat Mayonnaise

Viktoria Olsson [1,*], Andreas Håkansson [1], Jeanette Purhagen [2,3] and Karin Wendin [1,4]

1. Research Environment MEAL, Faculty of Natural Science, Kristianstad University, SE-291 88 Kristianstad, Sweden; andreas.hakansson@hkr.se (A.H.); karin.wendin@hkr.se (K.W.)
2. Perten Instruments AB, SE-254 66 Helsingborg, Sweden; jpurhagen@perten.com
3. Department of Food Technology, Engineering and Nutrition, Lund University, SE-221 00 Lund, Sweden
4. Department of Food Science, University of Copenhagen, DK-1958 Frederiksberg, Denmark
* Correspondence: viktoria.olsson@hkr.se; Tel.: +46-442-503-817

Received: 13 November 2017; Accepted: 12 January 2018; Published: 17 January 2018

Abstract: Varying processing conditions can strongly affect the microstructure of mayonnaise, opening up new applications for the creation of products tailored to meet different consumer preferences. The aim of the study was to evaluate the effect of emulsification intensity on sensory and instrumental characteristics of full-fat mayonnaise. Mayonnaise, based on a standard recipe, was processed at low and high emulsification intensities, with selected sensory and instrumental properties then evaluated using an analytical panel and a back extrusion method. The evaluation also included a commercial reference mayonnaise. The overall effects of a higher emulsification intensity on the sensory and instrumental characteristics of full-fat mayonnaise were limited. However, texture was affected, with a more intense emulsification resulting in a firmer mayonnaise according to both back extrusion data and the analytical sensory panel. Appearance, taste and flavor attributes were not affected by processing.

Keywords: mayonnaise; emulsification; sensory evaluation; texture; processing

1. Introduction

Mayonnaise is an oil-in-water emulsion stabilized by egg yolk and has been produced commercially for more than one hundred years [1]. Traditional mayonnaise is produced in a batch process by slowly adding the oil to the water phase under vigorous mixing, thereby creating an emulsion [2]. Industrially, mixing is achieved using high-intensity rotor-stator mixers, also referred to as high-shear mixers [3]. Although the taste and texture of mayonnaise is appreciated by many consumers, local markets often value different sensory properties. Therefore, as it is known that production techniques such as mixing/homogenization may have a considerable effect on the final product structure [1,4], better knowledge of how processing conditions affect the sensory and instrumental properties of the emulsion could help cater for such varying consumer preferences.

Due to a high oil content, mayonnaise exhibits a semisolid and viscoelastic behavior that influences its particular rheological properties, which in turn contribute to the perceived texture and flavor of the product [5]. In this context, texture is defined as the sensory perception of the structure of a food [6]. According to van Aken et al. [7], the rheological properties of a food product are very important for the perception of a creamy mouthfeel, although other authors have stressed that a variety of aspects may also play a role. For example, the oil droplet size is another parameter of interest due to its ability to influence product appearance, texture, and flavor profile [8].

One way in which the texture of mayonnaise is perceived by the consumer is through its processing and breakdown in the mouth (intra orally) before it is swallowed. In fact, most sensations associated with food texture occur only when the food is manipulated, deformed, or moved across the receptors in the mouth [4]. Through texture analysis, it is possible to choose a compression technique similar to that performed by the mouth, and then measure the behavior of the food using this technique. Such tests are valuable since they can confirm various textural properties, including the creaminess of mayonnaise.

Texture is also perceived outside the mouth (extra orally). Before the food item enters the mouth, visual cues related to the item's appearance provide information regarding its texture, while additional information can also be obtained by handling the food, e.g., by stirring, spooning, and cutting [4].

The emulsification taking place when mayonnaise is formed in rotor-stator mixers is relatively well understood, and proceeds via hydrodynamic interactions between the dispersed phase and the fluid in the rotor-stator region. Experiments suggest that the dispersed phase is predominantly broken up by turbulent viscous stresses [9]. The diameter of the emulsion drops, U, is produced in the rotor-stator mixer scales with the rotor tip-speed, determined by

$$U = \pi ND \qquad (1)$$

and decreases according to the power-law function [9,10]. In Equation (1), N is the rotor speed and D is the rotor diameter. Drop size also decreases with processing time, and scales with the average number of passages, p, through the rotor-stator region [9], which is written as

$$p = t\frac{Q}{V} \qquad (2)$$

where t is processing time, V is the fluid volume, and Q is the flow through the stator screen of the mixer [11,12], expressed as follows:

$$Q = N_Q ND^3 \qquad (3)$$

where N_Q is a mixer-specific design constant. However, the dynamics of the process is very slow and the droplet size continues to decrease after the emulsion has been processed for more than the equivalent of an average of 100 passages through the rotor-stator region [9].

The effect of processing conditions on the sensory response of mayonnaise is not as well understood as the effect on emulsion drop diameters. Furthermore, studies in which the rheological properties of mayonnaise have been related to perceived texture have predominantly focused on low-fat mayonnaises with oil concentrations ranging from 15–30% [13], thus creating a knowledge gap with regard to how full-fat (~75–80%) mayonnaises are affected.

It has been shown that fat content has a significant effect on perceived thickness and fattiness, with a higher fat content yielding a higher perception of both qualities. However, increased emulsification intensity, which produces smaller droplets, has the opposite effect and has also been shown to affect the perceived sweetness and whiteness of mayonnaise with added aromas [14]. Taste, flavor and textural attributes are also of interest in mayonnaise without added aromas.

The aim of this study was to evaluate the effect of emulsification intensity on the sensory and instrumental characteristics of full-fat mayonnaise.

2. Materials and Methods

2.1. Mayonnaise Ingredients and Processing

The ingredients in the experimental mayonnaise are presented in Table 1 and were assembled according to the following protocol: All ingredients, with the exception of the rapeseed oil and vinegar, were weighed into a 500 mL dispersing vessel (Kinematica, Luzern, Switzerland) and allowed to rest for adjustment to room temperature. The dispersing aggregate PT-DA 20/2 EC-E192 (Kinematica, Luzern, Switzerland) of the rotor-stator mixer (Kinematica Polytron, PT 2500 E, Luzern, Switzerland) was adjusted in the vessel and the mixture processed for 30 s at 6000 rpm. The rapeseed oil was then added, initially dropwise, and the mixture processed at 6000 rpm until all oil was emulsified. Following this stage, the vinegar was added and the mayonnaise mixed for an additional 30 s. Thereafter, one batch (High emulsification intensity) was processed at 9000 rpm (corresponding to a rotor tip-speed of 7.1 m/s) for 6 min, and a second batch (Low emulsification intensity) at 6000 rpm (rotor tip-speed 4.7 m/s) for 9.2 min. These processing times were chosen to achieve the same number of average rotor-stator passages for each rotor speed, i.e., equal p for different N (see Equations (1)–(3)). This procedure was repeated 3 times per processing condition for both batches (High and Low) and pooled in order to obtain enough mayonnaise for sensory and instrumental texture analysis. A commercial mayonnaise (Äkta Majonnäs, Findus), produced in Sweden, was also included as reference; this mayonnaise contained 81% (w/v) rapeseed oil and 4.6% (w/v) egg yolk (Table 2). The rationale for including a commercial reference was dual-fold; not only did it provide an indication of the sensory characteristics of the experimental mayonnaise in comparison to those of a commercial mayonnaise sold on the Swedish retail market, but it also served as a control when comparing the results of the present panel with those of other analytical panels.

Table 1. List of ingredients in the experimental mayonnaise.

Ingredient	Weight (g)	w/v (%)
Rapeseed oil	321.6	81.2
Egg yolk	34.1	8.6
Water	23.3	5.9
Mustard	10	2.5
Vinegar (acetic acid 12%)	4.8	1.2
Salt	1.2	0.3
Sugar	1.2	0.3

Table 2. List of ingredients in the commercial reference sample.

Ingredient	w/v (%)
Rapeseed oil	81
Vinegar	
Egg yolk	4.6
Mustard seeds	
Sugar	
Salt	0.7
White pepper	
Thickener (E401, E412, E417)	
Cayenne pepper	
Preservatives (E211)	
Colorant, beta-carotene	

2.2. Sensory Evaluation

Sensory evaluation and training were carried out over a period of three days by an external panel (Kristianstad University, Kristianstad, Sweden) of ten assessors, who were selected and trained according to the following guidelines: ISO 3972, ISO8586-1, and ISO8586-2. The sensory laboratory was designed according to ISO 8589 and sensory analysis performed using sensory descriptive analysis [15]. Across two training sessions lasting approximately 2 h each, the panel developed descriptions of the perceived sensory attributes of the products, generating a set of attributes and developing a consensus regarding the evaluation of each attribute (Table 3). Reference materials were used in training for selected attributes such as yellow color, acid taste, and egg flavor.

Table 3. Sensory attributes and definitions established by the panel.

Category	Attribute	Definition
Appearance	Shiny	Degree of shininess
	Yellow	Gradation from a weak to a strong tone of (vanilla) yellow
Texture (extra-oral)	Adhesiveness to spoon	Amount of mayonnaise remaining on the spoon when held vertically
	Firmness	Degree of resistance when stirring with a spoon
Texture (intra-oral)	Fatty mouthfeel	Graded from a little to a high grade of perceived fattiness
	Creaminess	Degree of creaminess; yoghurt used as reference
Taste	Acidity	Taste of sourness; vinegar and lemon used as reference
	Sweetness	The pure taste of sucrose; no reference used, evaluation relied on individual recollection of sweet taste
	Saltiness	The pure taste of sodium chloride; no reference used, evaluation relied on individual recollection of salty taste
Flavor	Egg flavor	Sulfur, boiled egg; boiled eggs used as reference
	Total flavor	The total intensity of taste and flavor

Product evaluations were performed individually, in isolated booths. Samples (20 g) of mayonnaise were served on coded, disposable plastic dishes and handled using a plastic spoon. The serving temperature was controlled by leaving the samples at room temperature for 10 min prior to serving. Panelists were instructed to rinse their mouths with still or carbonated water after each sample, and were also provided with fresh cucumber, apple, and soft white bread for further palate cleansing. Samples were coded with three-digit codes and served in a randomized order. The panelists then evaluated the perceived intensities by hand on a continuous 100 mm line-scale labeled "low intensity" at 10 mm and "high intensity" at 90 mm. During one evaluation session, lasting 60 min, the panelists evaluated duplicates of each product, with the intensity ratings then translated into numbers.

2.3. Instrumental Texture Analysis

Texture measurements were performed in duplicate, via a back extrusion method using a TVT-300XP analyzer (Perten Instruments AB, Stockholm, Sweden) equipped with a 7 kg load cell and a back extrusion set consisting of a sample container (50 mm diameter) and a compression plate (40 mm diameter). The sampling distance was 20 mm, the test speed 1 mm/s, and the retraction speed 5 mm/s. Texture properties (Table 4) were measured using the TexCalc software (version 4.0.4.67).

Table 4. Texture properties measured.

Texture Property	Definition
Firmness (g)	The maximum compression force
Work of compression (g·mm)	The ability of the sample to flow around the probe (see Figure 1)
Stickiness (g)	The maximum (negative) force recorded during the withdrawal phase.
Adhesiveness (g·mm)	The work required to withdraw the probe through the sample (see Figure 1)
Gradient 1 (g/mm)	The gradient of the first third of the compression distance
Gradient 2 (g/mm)	The gradient of the second third of the compression distance

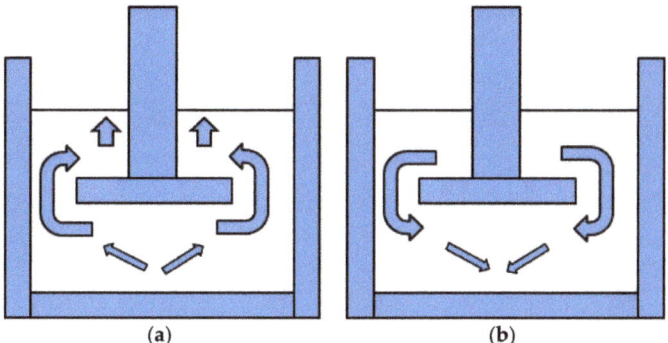

Figure 1. Movement of sample: (**a**) work of compression; (**b**) adhesiveness.

2.4. Statistical Methods

From the obtained data, mean values and standard deviations were calculated both for sensory and instrumental analysis. The sensory data were further subjected to a three-way analysis of variance (ANOVA), with samples panelists and replicates as fixed effects. The instrumental data were subjected to a one-way analysis of variance (ANOVA). Significant differences ($p < 0.05$) between samples were calculated via the Bonferroni's pairwise comparison test (Panel Check, v. 1.4.2, https://sourceforge.net/projects/sensorytool, Nofima, Norway).

Pearson correlations between instrumental and sensory data were calculated using Excel 2013 (Office for Windows), and Principal Component Analysis (PCA) performed using Panel Check (v. 1.4.2).

Panel performance was checked by calculating p- and MSE-values and then plotting these values in p-MSE-diagrams (Panel Check, v. 1.4.2).

3. Results

3.1. Sensory Evaluation

Panelist performance was found to be reliable based on the calculation of p-MSE-values showing low p- and MSE-values for all panelists. No significant effects were obtained due to replicates, with the exception of the attribute saltiness for which there was no significant difference between products.

Mixing intensity affected extra-oral texture attributes, i.e., those obtained by handling the mayonnaise through stirring and spooning (Table 5). The higher mixing intensity (rotor tip-speed 7.1 m/s) led to a significantly firmer, more viscous mayonnaise compared to the lower mixing intensity (rotor tip-speed 4.7 m/s) when handled by spoon. The results also indicated that the higher emulsification intensity produced a mayonnaise perceived as more creamy, although not to a statistically significant level.

Neither appearance nor taste or flavor attributes were affected by emulsion intensity (Table 5). The commercial, reference mayonnaise stood out in the sensory evaluation as having a more pronounced yellow color and a firmer, more creamy texture when assessed both extra- and intra-orally. The panel also perceived a more intense acidic flavor and a more intense total flavor in the commercial reference mayonnaise as compared to those prepared for the study (Figure 2).

Table 5. Sensory evaluation of the intensity of selected attributes, comparing the experimental mayonnaises and the commercial reference. Different letters in the same row indicate significant differences at $p \leq 0.05$.

Sensory Attribute	High Emulsification Intensity	Low Emulsification Intensity	Commercial Reference Mayonnaise
Shiny	62 ± 12 [a]	65 ± 9 [a]	70 ± 11 [a]
Yellow	43 ± 11 [a]	41 ± 12 [a]	84 ± 8 [b]
Adhesiveness to spoon	40 ± 17 [a]	58 ± 19 [b]	28 ± 14 [c]
Firmness	54 ± 9 [a]	47 ± 11 [b]	70 ± 12 [c]
Fatty mouthfeel	58 ± 17 [a]	57 ± 14 [a]	58 ± 14 [a]
Creaminess	60 ± 9 [a]	56 ± 13 [a]	70 ± 11 [b]
Acidity	34 ± 15 [a]	33 ± 11 [a]	55 ± 14 [b]
Sweetness	29 ± 10 [a]	28 ± 9 [a]	24 ± 9 [a]
Saltiness	23 ± 7 [a]	23 ± 7 [a]	28 ± 11 [a]
Egg flavor	32 ± 10 [a]	32 ± 9 [a]	35 ± 7 [a]
Total flavor	45 ± 10 [a]	43 ± 9 [a]	61 ± 13 [b]

Different letters in the same row indicate significant differences at $p \leq 0.05$.

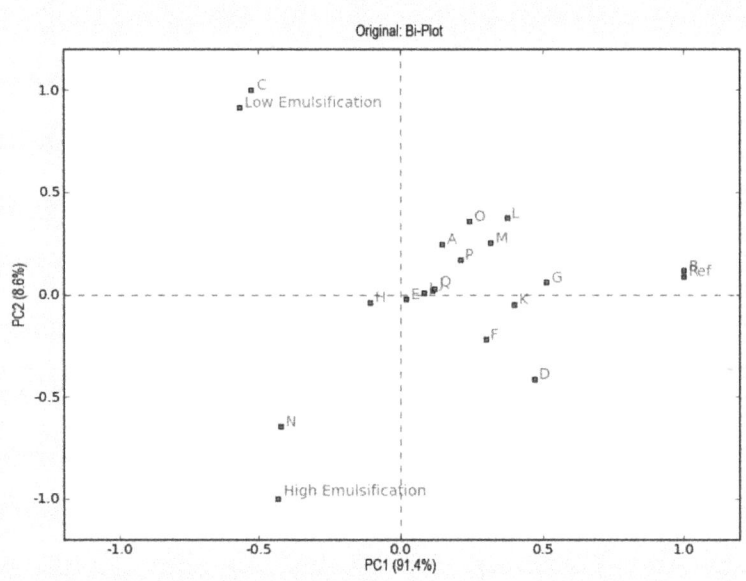

Figure 2. Principal component analysis (PCA) illustrating the sensory and instrumental characteristics of the experimental and commercial reference (Ref.) mayonnaises. The PCA plot shows 100% of the explained variance, meaning that the total variance is explained by two dimensions. A: Shiny; B: Yellow; C: Adhesiveness Spoon; D: Firmness; E: Fatty Mouthfeel; F: Creaminess; G: Acidity; H: Sweetness; I: Egg flavor; J: Saltiness; K: Total Flavour; L: Inst Firmness; M: Inst Compression; N: Inst Stickiness; O: Inst Adhesiveness: P: Inst Gradient 1; Q: Inst Gradient 2.

3.2. Instrumental Texture Analysis

All samples exhibited the same behavior during back extrusion measurements, but to different extents, as shown in Figure 3.

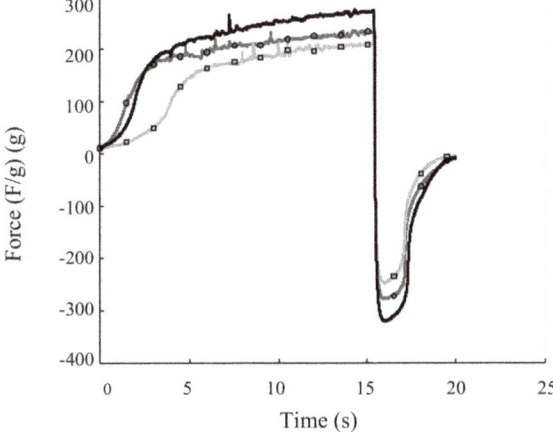

Figure 3. Graphs of back extrusion measurements. The black line represents the commercial reference mayonnaise, the gray line with circles the high emulsification intensity mayonnaise, and the gray line with squares the low emulsification intensity mayonnaise.

The commercial reference sample was found to be the most firm, sticky and adhesive mayonnaise, followed by the mayonnaise produced with a high emulsification intensity (Table 6).

Table 6. Instrumental texture analysis as performed via a back extrusion method. Different letters in the same row indicate significant differences at $p \leq 0.05$.

	High Emulsification Intensity	Low Emulsification Intensity	Commercial Reference Mayonnaise
Firmness (g)	252 ± 14 [a]	230 ± 15 [a]	292 ± 11 [b]
Stickiness (g)	−301 ± 22 [a]	−269 ± 19 [a]	−349 ± 23 [b]
Adhesiveness (J)	40 ± 3 [a]	35 ± 5 [a]	48 ± 3 [b]
Gradient 1 (g/mm)	29 ± 1 [b]	27 ± 1 [a]	37 ± 11 [a,b]
Gradient 2 (g/mm)	5 ± 2 [a]	5 ± 1 [a]	10 ± 3 [a]

Different letters in the same row indicate significant differences at $p \leq 0.05$.

Consequently, both the sensory and instrumental data imply that a higher emulsification intensity results in a firmer full-fat mayonnaise. When correlating the instrumental data to the sensory data the correlations were generally high ($r \geq 0.9$), primarily reflecting the pronounced difference between (i) the experimental mayonnaises (high and low emulsification intensity) and (ii) the commercial reference sample.

4. Discussion

A higher emulsification intensity affects the microstructure of mayonnaise by decreasing the droplet size [9,10], thus potentially affecting sensory traits such as texture [13], color [14], and flavor [8]. However, our results revealed no significant difference between the experimental samples with regard to color, taste, or flavor. As smaller particles increase light scattering, a reduced droplet size leads to a whiter mayonnaise, a phenomenon that has been illustrated in cream cheese, in which homogenization was found to lower the intensity of the yellow color [16]. In theory, flavor release decreases with increasing droplet size, as it takes longer for flavor molecules to diffuse out of a larger droplet. However, polar and non-polar flavor molecules behave differently in this respect, and the influence of droplet size on the rate of flavor release depends on the nature of the system [8]. In the study conducted by Wendin, Langton, Caous and Hall [16], smaller droplet sizes in cream cheese resulted in a shorter duration of the dynamic sensation of "fat-creamy".

The samples showed significant textural differences linked to the intensity of emulsification, with a more intense emulsification producing higher firmness and creaminess, as well as a decrease in adhesiveness to the spoon when handled. The textural attributes of mayonnaise can be explained by the elastic parameters of dynamic viscoelasticity (G′). This property is strongly related to particle size at 10% cumulative volume, which is in turn negatively correlated with sensory attributes including hardness, fracturability, and adhesiveness [13]. The perception of texture is a complex process involving the senses of vision, hearing, somesthesis, and kinesthesis [17]. Neurologically, texture perception results from the interaction of sensory and motor components of the peripheral nervous system with the central nervous system. Initially, the sight and extra-oral manipulation of food, e.g., through using a spoon, sets up sensory expectations regarding texture [18]. Then, once the food is put into the mouth, texture perception is a dynamic process, as the physical properties of foods change continuously when manipulated intra-orally. In this respect it would be interesting to further examine the question of how well texture attributes that are perceived extra-orally correlate with perceived oral-somatosensory attributes.

In the present study, the emulsion drop size distribution was not measured. However, previous investigations have shown that the scaling of drop-diameter averages with rotor tip-speed is highly predictable [9,10]. Using this previously established scaling, the higher emulsification intensity (rotor tip-speed 7.1 m/s compared to 4.7 m/s) corresponds to an expected reduction in the average oil drop diameter by a factor of two [9]. This is a rather substantial reduction that was expected to lead to quality differences with regard to the appearance, texture and flavor of the product. However, only extra-oral textural attributes were affected to a degree that could be perceived by the sensory panel.

Our findings, that a more intense emulsification and hence a decreased oil drop diameter produces a firmer mayonnaise, compare well with earlier results regarding the effect of microstructure on food emulsions [13,16]. The instrumental texture analysis data support the theory that a decreased droplet size leads to textural alterations, resulting in a more firm and adhesive mayonnaise. These findings may be helpful for the control and prediction of mayonnaise texture using processing conditions rather than more common approaches such as adding texture modifiers, which in the age of growing consumer preference for "clean labels" are unwanted in many products. Understanding the microstructural changes that occur during processing and the role of different mayonnaise ingredients will allow for better control of product structure and, ultimately, the manipulation and regulation of product texture [17].

The study conducted by Maruyama, Sakashita, Hagura and Suzuki [13] is just one among many to report that temperature during preparation may affect the physical properties of mayonnaise. Thus, if the aim is to obtain reproducible results, a consistent temperature is essential. In the present study, the ingredients were left to adjust to room temperature at approximately 20 °C, with the mayonnaise thereafter prepared at the same temperature. Compared to industrial preconditions, in which emulsification is commonly performed under cooling, the temperature in this study was high and control was inadequate, which might have influenced the results. Since emulsion formation is controlled by viscous drop breakup [9], a high temperature at the onset of emulsification will decrease the viscosity of the emulsion, reducing the viscous shear forces and thus resulting in larger drop sizes.

5. Conclusions

The effects of a higher emulsification intensity, corresponding to an expected reduction in the average oil droplet diameter by a factor of two, on the sensory and instrumental characteristics of full-fat mayonnaise were limited. Perceived and instrumentally measured texture was affected, with a more intense emulsification resulting in a firmer mayonnaise, as measured via back extrusion and by an analytical sensory panel. However, appearance, taste and flavor attributes were not affected by processing.

Acknowledgments: We hereby acknowledge the valuable technical assistance provided by Sarah Forsberg and Therése Svensson. Funding: This work was supported by Kristianstad University.

Author Contributions: Viktoria Olsson, Andreas Håkansson and Karin Wendin conceived and designed the experiments; Viktoria Olsson and Jeanette Purhagen performed the experiments; Karin Wendin and Andreas Håkansson analyzed the data; Viktoria Olsson, Andreas Håkansson, Jeanette Purhagen and Karin Wendin wrote the paper.

Conflicts of Interest: The authors declare no conflict of interest.

References

1. Harrison, L.; Cunningham, F. Factors influencing the quality of mayonnaise: A review. *J. Food Qual.* **1985**, *8*, 1–20. [CrossRef]
2. Depree, J.A.; Savage, G.P. Physical and flavour stability of mayonnaise. *Trends Food Sci. Technol.* **2001**, *12*, 157–163. [CrossRef]
3. Bengoechea, C.; Lopez, M.L.; Cordobes, F.; Guerrero, A. Influence of semicontinuous processing on the rheology and droplet size distribution of mayonnaise-like emulsions. *Food Sci. Technol. Int.* **2009**, *15*, 367–373. [CrossRef]
4. Engelen, L.; Van Der Bilt, A. Oral physiology and texture perception of semisolids. *J. Texture Stud.* **2008**, *39*, 83–113. [CrossRef]
5. Stern, P.; Valentova, H.; Pokorny, J. Rheological properties and sensory texture of mayonnaise. *Eur. J. Lipid Sci. Technol.* **2001**, *103*, 23–28. [CrossRef]
6. Weenen, H.; Van Gemert, L.J.; Van Doorn, J.M.; Dijksterhuis, G.B.; De Wijk, R.A. Texture and mouthfeel of semisolid foods: Commercial mayonnaises, dressings, custard desserts and warm sauces. *J. Texture Stud.* **2003**, *34*, 159–179. [CrossRef]
7. Van Aken, G.A.; Vingerhoeds, M.H.; de Wijk, R.A. Textural perception of liquid emulsions: Role of oil content, oil viscosity and emulsion viscosity. *Food Hydrocoll.* **2011**, *25*, 789–796. [CrossRef]
8. McClements, D.J. *Food Emulsions: Principles, Practices, and Techniques*, 3rd ed.; CRC Press: Boca Raton, FL, USA, 2016.
9. Håkansson, A.; Chaudhry, Z.; Innings, F. Model emulsions to study the mechanism of industrial mayonnaise emulsification. *Food Bioprod. Process.* **2016**, *98*, 189–195. [CrossRef]
10. Adler-Nissen, J.; Mason, S.L.; Jacobsen, C. Apparatus for emulsion production in small scale and under controlled shear conditions. *Food Bioprod. Process.* **2004**, *82*, 311–319. [CrossRef]
11. Cooke, M.; Rodgers, T.L.; Kowalski, A.J. Power consumption characteristics of an in-line silverson high shear mixer. *AIChE J.* **2012**, *58*, 1683–1692. [CrossRef]
12. Mortensen, H.H.; Innings, F.; Håkansson, A. The effect of stator design on flowrate and velocity fields in a rotor-stator mixer-an experimental investigation. *Chem. Eng. Res. Des.* **2017**, *121*, 245–254. [CrossRef]
13. Maruyama, K.; Sakashita, T.; Hagura, Y.; Suzuki, K. Relationship between rheology, particle size and texture of mayonnaise. *Food Sci. Technol. Res.* **2007**, *13*, 1–6. [CrossRef]
14. Wendin, K.; Ellekjaer, M.R.; Solheim, R. Fat content and homogenization effects on flavour and texture of mayonnaise with added aroma. *LWT-Food Sci. Technol.* **1999**, *32*, 377–383. [CrossRef]
15. Stone, H.; Sidel, J.L. *Sensory Evaluation Practices*, 3rd ed.; Academic Press: San Diego, CA, USA, 2004.
16. Wendin, K.; Langton, M.; Caous, L.; Hall, C. Dynamic analyses of sensory and microstructural properties of cream cheese. *Food Chem.* **2000**, *71*, 363–378. [CrossRef]
17. Wilkinson, C.; Dijksterhuis, G.B.; Minekus, M. From food structure to texture. *Trends Food Sci. Technol.* **2000**, *11*, 442–450. [CrossRef]
18. Spence, C.; Hobkinson, C.; Gallace, A.; Fiszman, B.P. A touch of gastronomy. *Flavour* **2013**, *2*, 14. [CrossRef]

© 2018 by the authors. Licensee MDPI, Basel, Switzerland. This article is an open access article distributed under the terms and conditions of the Creative Commons Attribution (CC BY) license (http://creativecommons.org/licenses/by/4.0/).

Article

Sensory Profile and Acceptability of HydroSOStainable Almonds

Leontina Lipan [1], Marina Cano-Lamadrid [1], Mireia Corell [2,3], Esther Sendra [1], Francisca Hernández [4], Laura Stan [5], Dan Cristian Vodnar [5], Laura Vázquez-Araújo [6,7] and Ángel A. Carbonell-Barrachina [1,*]

1. Department of Agro-Food Technology, Escuela Politécnica Superior de Orihuela, Universidad Miguel Hernández de Elche, Carretera de Beniel, km 3.2, 03312 Orihuela, Alicante, Spain; leontina.lipan@goumh.umh.es (L.L.); marina.cano.umh@gmail.com (M.C.-L.); esther.sendra@umh.es (E.S.)
2. Departamento de Ciencias Agroforestales, ETSIA, Universidad de Sevilla, Carretera de Utrera, km 1, 41013 Sevilla, Spain; mcorell@us.es
3. Unidad Asociada al CSIC de Uso sostenible del suelo y el agua en la agricultura (US-IRNAS), Crta de Utrera km 1, 41013 Sevilla, Spain
4. Department of Plant Science and Microbiology, Escuela Politécnica Superior de Orihuela, Universidad Miguel Hernández de Elche, Carretera de Beniel, km 3.2, 03312 Orihuela, Alicante, Spain; francisca.hernandez@umh.es
5. Faculty of Food Science and Technology, University of Agricultural Sciences and Veterinary Medicine Cluj-Napoca, 400372 Cluj-Napoca, 3-5 Manastur Street, Romania; laurastan@usamvcluj.ro (L.S.); dan.vodnar@usamvcluj.ro (D.C.V.)
6. Technological Center in Gastronomy, BCC Innovation (Basque Culinary Center Research and Innovation Center), Juan Avelino Barriola 101, 20009 Donostia-San Sebastián, Gipuzkoa, Spain; lvazquez@bculinary.com
7. Basque Culinary Center, Mondragon Unibersitatea, Juan Avelino Barriola 101, 20009 Donostia-San Sebastián, Gipuzkoa, Spain
* Correspondence: angel.carbonell@umh.es; Tel.: +34-670-345-050

Received: 25 December 2018; Accepted: 8 February 2019; Published: 12 February 2019

Abstract: Fresh water availability is considered highly risky because it is a finite resource, and a deficiency in water leads to numerous economic and environmental issues. Agriculture is one of the main consumers of fresh water in practices such as irrigation and fertilization. In this context, the main objectives of this study were (i) to determine the descriptive sensory profiles of four almond types grown using different irrigation strategies and (ii) to study their acceptance in a cross-cultural study (Romania and Spain). Consumers' willingness to pay for hydroSOS almonds was also evaluated. The four irrigation strategies evaluated were a control sample, two samples grown under regulated deficit irrigation strategies (RDI), and a sample grown under a sustained deficit irrigation strategy (SDI). The main conclusion was that neither descriptive nor affective sensory results showed significant differences among treatments. These findings should encourage farmers to reduce their water usage by demonstrating that sensory quality was not significantly affected by any of the studied treatments, compared to the control. Regarding willingness to pay, both Spanish and Romanian consumers were willing to pay a higher price for the hydroSOS almonds.

Keywords: cross-cultural affective test; descriptive sensory analysis; hydroSOStainable products; *Prunus dulcis*; willingness to pay

1. Introduction

Fresh water is a finite resource, and uncertainty regarding the remaining level of water for future generations has led the world to seek sustainability as a compulsory issue for future economic development and healthy ecosystems [1,2]. The World Economic Forum (WEF) placed water scarcity as the main global risk of the economy regarding impact, because a shortage of water means a stoppage of factories and food production, leading to the decline of the global economy [2]. The population growth drives to an augmentation in intensive food production that alters the environment due to greenhouse gas emissions, soil deterioration, and water stress [3].

Agriculture is one of the biggest consumers of fresh water, mainly due to the large volume necessary for irrigation (70–80% of the total) [1,4]. Opinions about irrigation in agriculture are divided about whether irrigation is necessary or not. Some believe water irrigation is required to produce enough food in the future due to world population growth, while others find irrigation agriculture wasteful because it creates "water-guzzling crops" [4]. For this reason, agriculture, particularly in the Mediterranean (semi-arid) region, must evaluate water use sustainability by implementing plans and irrigation strategies capable of reducing water irrigation but maintaining the quality of products [5].

Almonds are the major tree nut crop in the Mediterranean basin, which is defined by low rainfall and elevated evaporated demand during the almond growing cycle [6]. Although it is considered a drought-resistant crop, the almond tree (*Prunus dulcis*) needs irrigation to produce yield and profitability [6,7]. Numerous studies in fruits, such as almonds, olives, pistachio, apples, and grapes, have proven that fruit quality could be increased by controlling and reducing the amount of water irrigation [8–14]. Therefore, the development of deficit irrigation strategies (DI), such as regulated and sustained deficit irrigation, might be useful to increase the water productivity maintaining fruit quality.

Deficit irrigation strategies refer to the application of water below the crop evapotranspiration (ET: the combination between the evaporation losses from the soil and transpiration losses from the crop) requirements [4]. Regulated deficit irrigation (RDI) was developed to supervise vegetative vigor and consists of applying limited water during certain stages (in which plant is less sensitive to water stress) of the growing season. In the almond crop, the most recommended and less sensitive phenological period to apply water stress is the stage IV, which is contemporaneous with kernel filling and happens during the summer months of highest evaporative demand [5,6]. On the other hand, sustained deficit irrigation (SDI) is a strategy in which a uniform and reduced amount of water is applied to crops during all growing cycles, creating a progressive stress in plants throughout the season [6]. In this strategy, stress is produced by not entirely refilling the root zone when irrigated [7].

Sensory analysis techniques are essential to establish the quality of a product and to understand consumer preferences. Descriptive sensory analysis consists of detection and description, not only of quantitative but also qualitative sensory attributes of products. These attributes are of utmost importance to define a product, including its appearance, aroma, flavor, and texture [15]. On the other hand, affective tests are used to evaluate consumer preferences or acceptance responses to a product [15].

Society now expects the incorporation of an environmental sustainability plan [16]. In this context, consumers play an essential role because they demand and choose food products with specific characteristics, and nowadays they are concerned, not only about healthy diets but about environmental protection [3]. This has led to the development of the "environmentalism" phenomenon, which mostly relies on government agencies and non-governmental organizations caring about ecological issues [17]. These phenomena have made the consumer more conscious and interested in healthy, safe, and environmentally friendly food; consequently, the consumer has a greater willingness to pay for eco-friendly and hydroSOStainable (hydroSOS) products [17,18].

Under these circumstances, the aim of this study was (i) to determine the descriptive sensory profiles of four different almond types grown using different irrigation strategies (including hydroSOS samples) and (ii) to study their acceptance and consumers' willingness to pay in a cross-cultural study in Romania and Spain. Understanding consumers' preferences and their willingness to pay for hydroSOS almonds is vital for almond growers.

2. Materials and Methods

2.1. Irrigation Treatments

The almond cultivar used in the present study was "Vairo" and was grown on the commercial farm "La Florida" located in Dos Hermanas (Seville, Spain). The following four irrigation treatments were evaluated:

- T1 was full irrigation treatment using 433 ± 26 mm of applied water throughout the season with a stress integral of SI = 54.2. Trees were irrigated to assure the estimated crop ET, and thus represented the control.
- T2 were trees under regulated deficit irrigation (RDI) at optimum level (148 ± 24 mm; SI = 91.7). For irrigation scheduling, midday stem water potential (SWP) and maximum daily shrinkage (MDS) measurements were done. Then, in stage IV (kernel filling) of the almond growing cycle, the trees were irrigated when SWP was lower than −1.5 MPa or when MDS signal was above 1.75. The rest of the stages were irrigated to the SWP proposed by McCutchan and Shackel (1992) or MDS equal 1 [19].
- T3 trees were also irrigated under regulated deficit irrigation but in more severe conditions (103 ± 13 mm; SI = 94.9). Thus, the stage IV trees were irrigated when SWP was lower than −2 MPa or MSD signal above 2.75, and similar conditions as previously described for T2 were applied for the rest of the period.
- T4 trees were irrigated under sustained deficit irrigation (SDI) conditions (114 ± 13 mm; SI = 74.7). Water was applied gradually throughout the growing season.

In order to determine the accumulative effect of water deficit, water stress integral was calculated by using the following equation:

$$\text{SI} = \left| \sum (\min \Psi_{\text{stem}} - (-0.2)) \right| \times n \qquad (1)$$

In this expression, SI was the stress integral, min Ψ_{stem} was the average of minimum SWP, and n represented the day numbers interval.

The field study was conducted during 2017; in August, almonds were harvested, dried (below 5% moisture content), and delivered to University Miguel Hernández of Elche (Spain) facilities, where descriptive and affective studies were carried out. Almonds were also sent to University of Agricultural Sciences and Veterinary Medicine of Cluj-Napoca (Romania) to perform affective studies with Romanian consumers. Around 1.5 kg of almond kernels was needed for each treatment.

2.2. Descriptive Sensory Analysis

A trained panel with 10 highly trained panelists from the Food Quality and Safety Group (Miguel Hernández University of Elche, Orihuela, Alicante, Spain) conducted the descriptive analysis. Each panelist had more than 600 hours of experience with different types of food products. Although the panel had a vast experience in tasting almond and turrón (traditional Spanish dessert made basically of toasted almonds and honey), they had four orientation sessions for the almond tasting, where the panelists decided the final list of descriptors and reference products for each attribute. The reference and modified lexicon were the ones developed by Vázquez-Araújo et al. [20]. Table 1 shows the reference products used by the panelists for flavor and texture characterization. The almond color scale was developed using instrumental color measurements carried out with a Minolta Colorimeter CR-300 (Minolta, Osaka, Japan) in 400 almonds. The minimum, mean, and maximum values from instrumental color intensities were later converted into pantones with an online program Nix Color Sensor [21] and presented to the panelists as references. The ΔE shows the degree of total color change [22] and was calculated as,

$$\Delta E = [(L - L^*)^2 + (a - a^*)^2 + (b - b^*)^2]^{0.5} \qquad (2)$$

Table 1. Sensory attributes, reference materials, and their corresponding intensities, used for the descriptive analysis of almonds.

Descriptor	Definition	Reference [‡]	Intensity
		Appearance	
Color	The intensity of color from light to dark	$L^* = 51.3; a^* = 20.6; b^* = 38.8$	1.0
		$L^* = 51.3; a^* = 20.6; b^* = 38.8$	5.0
		$L^* = 51.3; a^* = 20.6; b^* = 38.8$	10.0
Size	The visual width of the of the almond from side to side	8–9 mm	1.0
		13–14 mm	5.0
		17–18 mm	9.0
Roughness	The number of hills and valleys perceived by the human eye on the almond surface (visual measured)	0%	1.0
		50%	5.0
		100%	10.0
		Basic Taste and Flavor	
Saltiness	The basic taste associated with a sodium chloride solution	0.15% NaCl	1.0
		0.25% NaCl	3.0
Sweetness	The basic taste associated with a sucrose solution	1% sucrose	3.0
		2% sucrose	5.0
Bitterness	The basic taste associated with a caffeine solution	0.01% caffeine	2.0
		0.02% caffeine	3.0
Astringency	A drying and puckering sensation on the mouth surface	0.03% alum	1.0
		Unripe dates	10.0
Overall nuts	Aromatics related to nuts in general	Mix of grinded Hacendado Nutget:hazelnut, 1:1	5.5
Almond ID	Aromatics reminiscent of almond	Marcona almonds	6.5
Benzaldehyde like	Artificial almond or cherry aromatics	Aroma: almond extract Dr. Oetker	10.0
		Flavor: bitter almond	10.0
Woody	The sweet, musty, dark, and dry aromatics associated with the tree bark	Hacendado walnuts	3.0
Aftertaste	Longevity of key attributes intensity after swallow the sample	30 s	1.0
		1 min	3.0
		1.5 min	6.0
		Texture	
Hardness	The force required to bite completely through the sample with molar teeth. Evaluated on the first bite down with the molars	Baby Bell light cheese	3.0
		Sugus chewy candy	6.0
		Hacendado almond	7.5
		Solano candy	10.0
Cohesiveness	The degree to which the sample deforms prior to breaking apart when compressed between molars	Hochland cheese slices	3.5
		Hacendado raisins	6.5
		Sugus chewy candy	8.0
Crispiness	The intensity of audible noise at first chew with molars	Nestlé cheerios	5.5
		Nestlé fitness	7.0
Fracturability	The force needed to break the almond. The evaluation was done with the molars after first chew	Nestlé cheerios	2.5
		Nestlé fitness	5.0
Adhesiveness	The effort needed to completely remove the sample from the teeth; measured after 5 chews	Kraft Miracle whip light dressing	4.5
		Marshmallow fluff	6.5
		Jif creamy peanut butter	8.5

[‡] Intensities are based on a 10-point numerical scale with 0.5 increments, where 0 means "none" and 10 means "extremely strong".

Almond roughness was visually measured and refers to the number of hills and valleys perceived by the human eye on the almond surface [23]. Almond size scale was prepared from the one used by Regulating Council of the Protected Geographical Indications of *Jijona and Turrón de Alicante* (RCPGIJTA) [24].

The texture attributes (Table 1) products were also analyzed using instrumental texture measurements. A texture analyzer (Stable Micro Systems, model TA-XT2i, Godalming, UK) was employed using a 30 kg load cell and a Volodkevich Bite Jaw HDP/VB probe (trigger was set at 15 g, test speed was 1 mm s^{-1} over a specified distance of 3 mm).

After the orientation session, each panelist received four samples corresponding to the different irrigation treatments, and three evaluations per sample were done. The samples were served in odor-free 30 mL covered plastic cup and randomly coded with three digits. Water and unsalted crackers were also provided in order to clean the palates among samples. The descriptive test was carried out in a special tasting room with individual booths (controlled temperature of 21 ± 1 °C and combined natural/artificial light), and ballot charts were used to collect panelists' evaluations. The samples were presented according to a randomized block design to avoid biases. A 0 to 10 numerical scale was used by the panelists to quantify the intensity of the almond attributes, where 0 represents none/no intensity and 10 extremely strong with a 0.5 increment.

2.3. Affective Sensory Analysis

Affective sensory analysis was carried out with 100 recruited consumers from Spain (S) and 100 from Romania (R), with a gender ratio of 50:50 in Spain and 60:50 women:men in Romania. The consumers' age range was 18–25 (S = 33%; R = 45%), 26–35 (S = 29%; R = 30%), 36–45 (S = 10%; R = 15%), and 45–60 (S = 29%; R = 10%). The recruitment process was conducted via e-mail and fliers. Demographic questions regarding gender, age, nut consumption frequency, allergies, intolerances, or diet restriction were also included in the questionnaire. Spanish to Romanian back-translation procedure was conducted to avoid major misunderstandings during the evaluation. All samples were served, and labeled with three digit codes, in the same manner as with the recipients as described above. Consumers were asked for global satisfaction degree using a 9-point hedonic scale (1 = dislike extremely and 9 = like extremely) for scoring and about attributes intensity using Just About Right (JAR) questions. Consumers were also asked to rank samples according to their preference and to check the reasons why they choose that sample as the best (due to the color, flavor sweetness, crunchiness, etc.) by using a question type Check All That Apply (CATA). Consumer interest in the label information (sustainable, bio, healthy, natural, product of Spain/Romania, etc.) using CATA question type was also analyzed. As described in the descriptive section, the affective tests were also carried out in special tasting rooms with individual booths and according to a randomized block design.

2.4. Consumer Willingness to Pay

The willingness to pay was carried out with 100 consumers form Spain and 100 consumers from Romania. Both Spanish and Romanian consumers were first given information about what the hydroSOStainable concept means, and later they were asked for their willingness to pay for hydroSOS almonds compared to the conventional ones. It was decided to inform consumers about hydrosustainability, because it was a relatively new concept and because previous studies have demonstrated that consumers need enough knowledge and access to precise information to prevent the receiving of fake feedback [3]. Without this previous basic information about the hydroSOStainable concept, consumers' responses and resulting conclusions with regard to hydroSOS almonds would be deeply speculative [25]. Later, they were given a price for conventional almonds of 2.60 €/200 g (the normal price for the Mercadona almonds; Mercadona is one of the most popular food supermarkets in the Mediterranean area of Spain) and the options: ≤€2.60, €3.10, €3.60, and >€3.60.

2.5. Statistical Analysis

Statistical analyses were performed by subjecting the data to two or three-way analysis of variance (ANOVA) and then to Tukey's multiple range test. A three-way ANOVA (factor 1: irrigation treatment; factor 2: session; and, factor 3: panelist) was carried out to demonstrate the panel consistency in the descriptive sensory analysis data, while two-way ANOVA (factor 1: irrigation treatment, and factor 2: country) was used for the affective sensory data [26]. Statistically significant differences were considered when $p < 0.05$, and were performed using XLSTAT Premium 2016 (Addinsoft, New York, NY, USA) and Statgraphics Plus (Version 3.1, Statistical Graphics Corp., Rockville, MA, USA).

Penalty analysis was also carried out to supply information about the possible improvement of samples, and for these analyses, JAR data were used [26]. Mean drops (penalties) versus the percentage of the consumers (providing each response in the mean drop plot) were graphically represented.

3. Results and Discussion

3.1. Descriptive Sensory Analysis

The descriptive sensory analysis was performed to evaluate whether significant differences among treatments were found. The descriptive results showed no statistically significant differences for session and panelist, and their two-way interaction demonstrated proper performance of the panel and the lack of effects of the parameters panelist and replication. Thus, only the effect of the parameter "irrigation treatment" is presented and discussed in the manuscript.

Table 2 shows the effect of the studied irrigation treatments on the main sensory descriptors of control and hydroSOS almonds. No significant differences were observed for 12 out of the 17 attributes used to describe the quality of almonds, while statistically significant differences were found for color, size, roughness, sweetness, and hardness.

Panelists found T2 samples having more intense color. This was supported by the instrumental color data, which showed significantly higher values for the a^* coordinate (T1 = 16.7 b; T2 = 17.4 a; T3 = 17.1 ab; T4 = 17.3 ab). Although the differences for AE color were below two units, and differences are difficult to perceive with the human eye [27], the highest values in the a^* (green-red coordinate) indicated that T2 almonds were more reddish. However, other authors obtained no statistically significant differences for this parameter in pistachio and olives [8,10,28]. However, other authors showed that total color increased for apricots and peaches under RDI [29,30].

With respect to the size attribute (Table 2), no significant differences were observed in instrumental size (mm) among treatments (T1 = 16.3; T2 = 16.2; T3 = 16.2; T4 = 16.2), but slightly significant differences were observed in descriptive analysis for T1 and T2, compared to a second group, T3 and T4. The results were partially similar to previous studies about hydroSOS pistachios, in which it was observed that there were no significant differences either for sensory or for instrumental size [10]. It is a general working hypothesis that deficit irrigation can lead to reduced yield, but the fruits produced will be of higher size. The problem of working with fruits, such as almonds, is that the heterogeneity of the fruits is so high that in some cases and parameters/attributes it can mask real differences due to the applied treatments. This is one reason, among others, for this hypothesis not being confirmed by real data. However, current field experiments are being repeated during three years to have more realistic and reliable data, but preliminary data is needed to be able to implement improvements in the experiment design and reach partial goals and objectives.

Although the roughness (Table 2) of T2 and T4 recorded the highest values, they were within the optimal values, which meant that water stress was correctly applied. Applying water stress at the wrong growing stages, for instance, stage III, will lead to very rough kernels, which are indicative that the water stress also reached the fruit and its turgor and moisture content were drastically limited [31].

An important finding was that T2 and T3 were the sweetest samples. An increment in sweetness was demonstrated in "Mollar de Elche" pomegranate cultivar growth under deficit irrigation conditions [32]. Sweetness is a key attribute in the sensory quality of almonds, and it is expected that increased sweetness intensity will be favorable for consumer satisfaction [33]. Table 2 shows texture results, and only significant differences were found for hardness; T1 almonds were slightly softer than those from the rest of treatments. However, no differences were found for the instrumental hardness (T1 = 73.8 N; T = 73.8 N; T3 = 72.8 N; T4 = 72.2 N). Other authors also showed higher values for both sensory and instrumental texture for DI samples in studies about pistachios and olives samples [8,10].

3.2. Affective Sensory Analysis

Table 3 showed that the overall and attribute specific satisfaction degree of both Spanish and Romanian consumers were not statistically affected by the irrigation strategies under evaluation, with the exception of T4 almonds causing a slightly higher satisfaction. In general, Romanians tend to score higher than Spanish consumers because of the fact that they are less used to consuming this nut. Other authors, in studies about olives under deficit irrigation conditions, also reported no significant differences among the treatments for affective sensory evaluation [9]. On the contrary, there are also plenty of works on olives, pistachio, peaches, and grapes, in which higher consumer acceptance for samples produced by deficit irrigation strategies, such as RDI (moderate level), were observed [8,10,34,35].

Table 2. Descriptive sensory analysis of raw almonds as affected by deficit irrigation. Scale used ranged from 0 = no intensity to 10 = extremely strong intensity.

Irrigation Treatments	Color	Size	Roughness	Saltiness	Sweetness	Bitterness	Astringency	Overall Nuts	Almonds ID	Benzaldehyde-like	Woody	Aftertaste	Hardness	Cohesiveness	Crispiness	Fracturability	Adhesiveness
ANOVA Test [†]	***	*	***	NS	*	NS	NS	NS	NS	NS	NS	NS	***	NS	NS	NS	NS
Tukey Multiple Range Test [‡]																	
T1	4.0 c	8.0 a	5.3 b	0.5	3.3 ab	0.6	0.5	5.6	5.9	0.4	1.5	5.4	4.5 b	3.0	3.3	2.1	6.7
T2	5.3 a	8.0 a	6.7 a	0.5	3.5 a	0.6	0.6	5.8	5.9	0.3	1.0	5.4	5.1 a	2.7	3.7	2.5	6.6
T3	4.2 c	7.7 b	5.3 b	0.5	3.5 a	0.6	0.6	5.9	6.2	0.3	1.8	6.1	5.6 a	3.3	4.0	2.1	6.5
T4	4.7 b	7.6 b	6.2 a	0.4	2.7 b	0.5	0.7	5.4	5.9	0.3	1.9	6.1	5.6 a	3.3	4.2	2.5	6.4

[†] NS = not significant at $p < 0.05$; *, **, and *** significant at $p < 0.05$, 0.01, and 0.001, respectively. [‡] Values (mean of 10 trained panelists) followed by the same letter, within the same column, were not significantly different ($p < 0.05$), according to Tukey's least significant difference test.

Table 3. Affective sensory analysis of raw almonds as affected by deficit irrigation and tested in two countries of Union Europe (Spain and Romania).

	Color	Size	Almond ID	Sweetness	Bitterness	Astringency	Firmness	Crispiness	Teeth Adhesion	Aftertaste	Overall
ANOVA [†]											
Irrigation	NS	NS	NS	NS	NS	NS	NS	NS	NS	*	NS
Country	***	*	***	***	NS	NS	***	***	**	***	***
Irrigation × Country	NS	NS	***	*	NS	NS	**	***	**	***	***
Tukey Multiple Range Test [‡]											
Irrigation											
T1	7.0	7.0	6.7	6.6	6.4	6.5	6.7	6.9	6.3	6.5 ab	6.6
T2	7.0	7.2	6.8	6.6	6.5	6.5	6.5	6.9	6.2	6.2 b	6.5
T3	7.2	7.1	7.0	6.7	6.5	6.6	6.9	7.1	6.5	6.6 ab	6.9
T4	7.0	7.2	7.0	6.8	6.4	6.5	6.7	7.0	6.7	6.8 a	7.0
Country											
Spain	6.8 b	6.9 b	6.3 b	6.3 b	6.5	6.6	6.3 b	6.4 b	6.1 b	6.0 b	6.3 b
Romania	7.2 a	7.3 a	7.1 a	6.9 a	6.5	6.5	6.9 a	7.3 a	6.6 a	6.7 a	7.0 a
Irrigation × Country											
Spain											
T1	6.7	6.9	6.2 b	6.2 c	6.6	6.6	6.2 ab	6.4 b	6.2 ab	6.1 bc	6.2 bc
T2	6.6	6.8	6.1 b	6.2 c	6.4	6.3	6.0 b	6.3 cd	5.9 b	5.6 c	6.0 c
T3	7.0	7.2	6.4 b	6.4 b	6.4	6.8	6.5 ab	6.6 abcd	6.2 ab	6.2 abc	6.5 abc
T4	6.8	6.9	6.5 ab	6.3 bc	6.4	6.6	6.4 ab	6.4 bcd	6.3 ab	6.3 abc	6.4 bc
Romania											
T1	7.2	7.1	7.0 ab	6.8 ab	6.3	6.5	6.9 ab	7.3 ab	6.4 ab	6.7 ab	6.8 abc
T2	7.1	7.5	7.1 a	6.7 ab	6.5	6.6	6.7 ab	7.2 abc	6.4 ab	6.4 abc	6.8 abc
T3	7.3	7.1	7.2 a	6.8 ab	6.5	6.5	7.1 a	7.4 a	6.7 ab	6.8 ab	7.0 ab
T4	7.1	7.4	7.3 a	7.0 a	6.5	6.5	6.9 ab	7.4 a	6.9 a	7.1 a	7.3 a

[†] NS = not significant at $p < 0.05$; *, **, and *** significant at $p < 0.05$, 0.01, and 0.001, respectively. [‡] Values (mean of 100 consumers) followed by the same letter, within the same column, were not significantly different ($p < 0.05$), according to Tukey's least significant difference test.

Figure 1 shows the sample preference order of Spanish (S) and Romanian consumers (R), and the main attributes controlling their preference. Sample T2 was chosen by consumers from both countries as the most liked sample, while T1 almonds were the least like ones (Figure 1a). Spanish consumers scored T4 higher than T3, while the contrary was observed from the Romanian consumers. T2 almonds were chosen as the best ones mainly due to their almond flavor, sweetness, and crispiness (Figure 1b), showing that sweetness was important in consumers' satisfaction, as hypothesized before in this study.

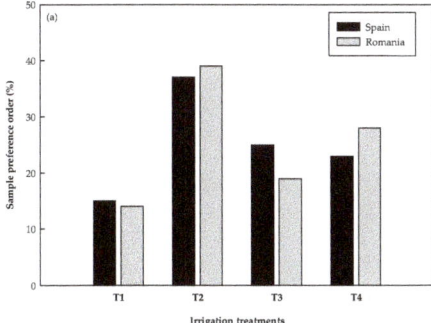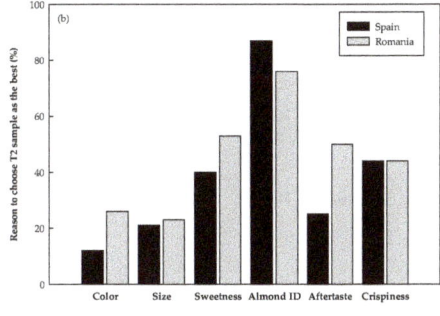

Figure 1. Purchase intent of Spanish and Romanian consumers regarding the studied almonds (**a**), and their reason to choose T2 almonds as the favorite ones (**b**).

Besides, sweetness (S = 29%; R = 64%), flavor (S = 77%; R = 65%), texture (S = 44%; R = 16%), and price (S = 44%; R = 60%) were the most checked parameters in the CATA questionnaire used when consumers were asked about their buying drivers. The most important word in a product label for 63% of the Spanish consumers was "product of Spain", followed by "healthy" (52%) and "natural" (48%), while Romanian consumers were more interested in "natural" (70%), "healthy" (67%), and "ecological" (31%). For both nationalities, the words "natural" and "healthy" seemed to play a key role in their buy decisions. Noguera et al. [18] also found the word "product of Spain" as the most important attribute for Spanish consumers concerning hydroSOS pistachio along with other expressions such as "rich in antioxidants" and "crunchy". The findings were associated with the consumer recognition of national products, health concerns, and composition. Regarding the word "sustainable", 44% of Spanish consumers and 29% of Romanians were interested in this word when purchasing a product. Although it is a relatively new concept [36], other authors also showed great interest of Spanish consumers for this word and concept but when studying other products, pistachios [18].

Penalty analysis, a very popular method in the food industry sector to help one interpret data from JAR questions, was conducted to understand the relationship between consumers' overall liking and the attribute intensity scores of the JAR questions [37,38]. The attributes with a large penalty and a high percentage of consumers, which were placed in the upper right quadrant of the plots, provided information about the most critical diagnostic issues of the product. On the other hand, the preferred attributes were usually located at the lower left quadrant of the plots [26]. The proportion of consumer's opinion plots and the mean penalty is shown in Figure 2 for all four treatments under study. All attributes having a negative impact on the sample liking, for at least 20% of the consumers and producing a drop of at least 1 point for liking, are the ones that might need to be improved. Bitterness was the only parameter susceptible to be improved, and consumers from both Spain and Romania agreed that this was especially true for T1 and T4 almonds.

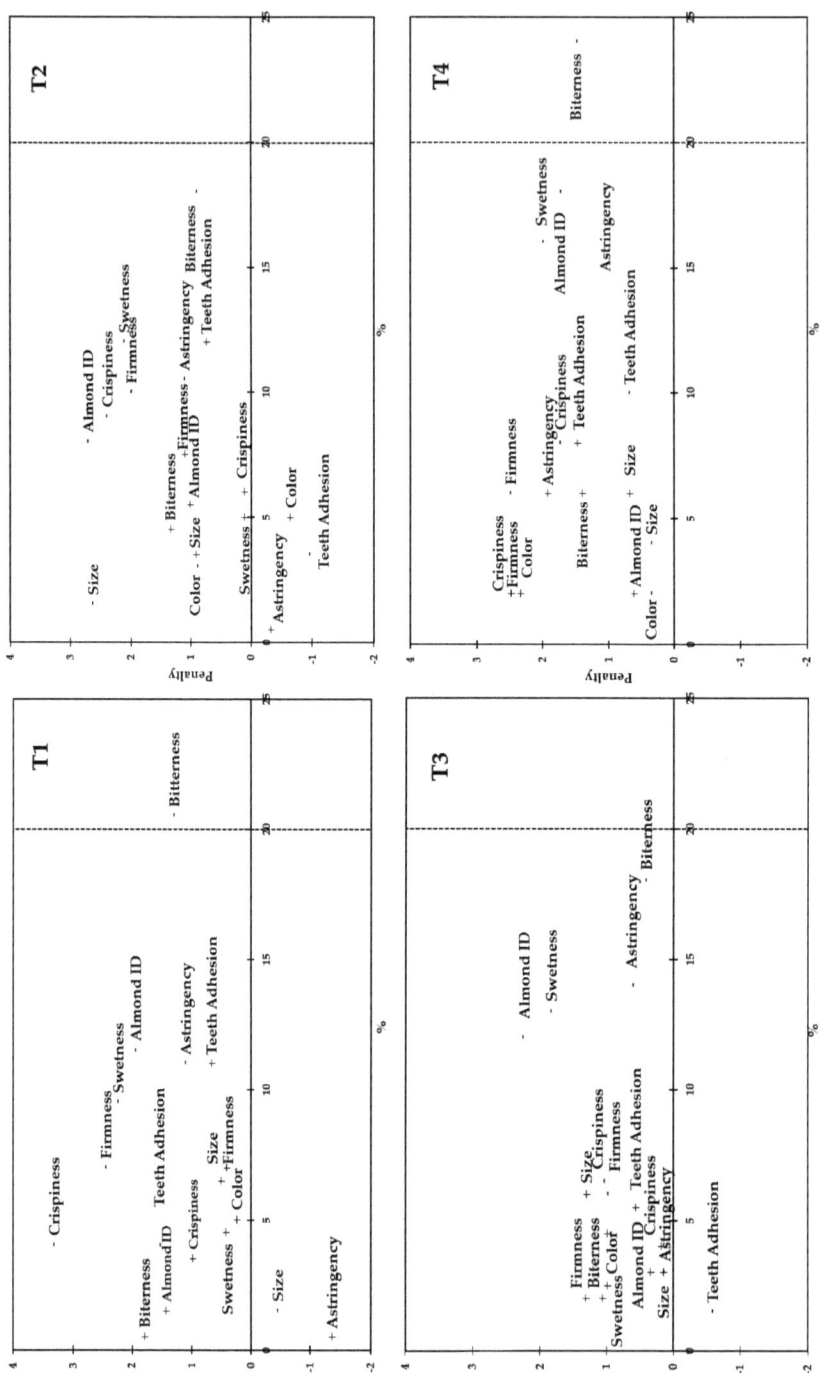

Figure 2. Penalty analysis of attributes intensities assessed by consumers (sample code indicated on the top right of each figure; "too low intensity" is indicated by the symbol "−", and "too high intensity" is indicated by the symbol "+").

3.3. Consumer Willingness to Pay

The Spanish and Romanian consumers were classified according to their willingness to pay for a bag of hydroSOS almonds compared to a bag of conventional almonds: (i) S = 23% and R = 31% were willing to pay less or the same price; (ii) S = 60% and R = 16% wanted to pay 0.50 € more; (iii) S = 13% and R = 24% wanted to pay 1.00 € more; and, finally, (iv) S = 4% and R = 29% wanted to pay more than 1.00 €. These findings agreed with Noguera et al. [18], who reported that Spanish consumers were also willing to pay an extra amount of money for hydroSOS pistachios [18].

Considering the almond Spanish production of ~ 190.000 t, the price received by the farmers for conventional shelled almonds was ~ 4.85 € kg^{-1} [39], and according to the previous data, an extra value of ~2 € kg^{-1} hydroSOS almonds can be expected. The possible economic increase by using deficit irrigation strategies could be ~40% with respect to conventional almonds. These gains might encourage farmers to invest in these novel sustainability tools, contributing to environmentally friendly agriculture.

4. Conclusions

The present study was the first to analyze the sensory properties of hidroSOStaninable almonds and consumers' (Romania and Spain) acceptance and willingness to pay. Although the consumer panels showed similar global and attribute-specific satisfaction degrees, the trained panelists were able to establish slight but significant differences in some key attributes, with T2 almonds showing intense red color, high size, and high intensity of both sweetness and hardness attributes. The penalty analysis also showed that bitterness, which was susceptible to be improved in other treatments, was correct in T2. Consumers are now aware of the importance of the environment and the need to optimize key resources, such as water. This awareness may explain consumers' willingness to pay a higher price for hydroSOS almonds, which will lead to higher incomes and benefits for farmers. These results lead us to conclude that controlling stress in almond trees with deficit irrigation strategies can increase water productivity and farmers' profits from producing of environmentally friendly products without significantly changing the sensory profile and the consumers' satisfaction.

Author Contributions: Conceptualization, Á.A.C.-B. and L.V.-A.; Methodology, L.L., D.C.V. and L.S.; Software, E.S.; Validation, Á.A.C.-B., F.H. and E.S.; Formal Analysis, L.L.; Investigation, L.L.; Resources, Á.A.C.-B.; Data Curation, L.L.; Writing—Original Draft Preparation, L.L.; Writing-Review & Editing, L.L. and M.C.-L.; Visualization, Á.A.C.-B.; Supervision, L.V.-A.; Project Administration, Á.A.C.-B.; Funding Acquisition, Á.A.C.-B.; Field Experiments, M.C.

Funding: The study has been funded (Spanish Ministry of Economy, Industry, and Competitiveness) through a coordinated research project (hydroSOS) including the Universidad Miguel Hernández de Elche (AGL2016-75794-C4-1-R, hydroSOS foods) and the Universidad de Sevilla (AGL2016-75794-C4-4-R). Marina Cano-Lamadrid was funded by a FPU grant from the Spanish Ministry of Education (FPU15/02158).

Conflicts of Interest: The authors declare no conflict of interest.

References

1. D'Ambrosio, E.; De Girolamo, A.M.; Rulli, M.C. Assessing sustainability of agriculture through water footprint analysis and in-stream monitoring activities. *J. Clean. Prod.* **2018**, *200*, 454–470. [CrossRef]
2. Hogeboom, R.J.; Kamphuis, I.; Hoekstra, A.Y. Water sustainability of investors: Development and application of an assessment framework. *J. Clean. Prod.* **2018**, *202*, 642–648. [CrossRef]
3. Lazzarini, G.A.; Visschers, V.H.M.; Siegrist, M. How to improve consumers' environmental sustainability judgements of foods. *J. Clean. Prod.* **2018**, *198*, 564–574. [CrossRef]
4. Fereres, E.; Soriano, M.A. Deficit irrigation for reducing agricultural water use. *J. Exp. Bot.* **2007**, *58*, 147–159. [CrossRef]

5. Lipan, L.; Sánchez-Rodríguez, L.; Collado González, J.; Sendra, E.; Burló, F.; Hernández, F.; Vodnar, D.-C.; Carbonell-Barrachina, A.A. Sustainability of the legal endowments of water in almond trees and a new generation of high quality hydrosustainable almonds—A review. *Bull. UASVM Food Sci Technol.* **2018**, *75*, 97–108. [CrossRef]
6. Egea, G.; Nortes, P.A.; Domingo, R.; Baille, A.; Pérez-Pastor, A.; González-Real, M.M. Almond agronomic response to long-term deficit irrigation applied since orchard establishment. *Irrig. Sci.* **2013**, *31*, 445–454. [CrossRef]
7. Goldhamer, D.A.; Viveros, M.; Salinas, M. Regulated deficit irrigation in almonds: Effects of variations in applied water and stress timing on yield and yield components. *Irrig. Sci.* **2005**, *24*, 101–114. [CrossRef]
8. Cano-Lamadrid, M.; Girón, I.F.; Pleite, R.; Burló, F.; Corell, M.; Moriana, A.; Carbonell-Barrachina, A.A. Quality attributes of table olives as affected by regulated deficit irrigation. *LWT Food Sci. Tech.* **2015**, *62*, 19–26. [CrossRef]
9. Cano-Lamadrid, M.; Hernández, F.; Corell, M.; Burló, F.; Legua, P.; Moriana, A.; Carbonell-Barrachina, A.A. Antioxidant capacity, fatty acids profile, and descriptive sensory analysis of table olives as affected by deficit irrigation. *J. Sci. Food Agric.* **2016**, *97*, 444–451. [CrossRef]
10. Carbonell-Barrachina, A.A.; Memmi, H.; Noguera-Artiaga, L.; Gijón-López Mdel, C.; Ciapa, R.; Pérez-López, D. Quality attributes of pistachio nuts as affected by rootstock and deficit irrigation. *J. Sci. Food Agric.* **2015**, *95*, 2866–2873. [CrossRef]
11. Egea, G.; González-Real, M.M.; Baille, A.; Nortes, P.A.; Sánchez-Bel, P.; Domingo, R. The effects of contrasted deficit irrigation strategies on the fruit growth and kernel quality of mature almond trees. *Agric. Water Manag.* **2009**, *96*, 1605–1614. [CrossRef]
12. Sánchez-Rodríguez, L.; Corell, M.; Hernández, F.; Sendra, E.; Moriana, A.; Carbonell-Barrachina, Á.A. Effect of Spanish-style processing on the quality attributes of HydroSOStainable green olives. *J. Sci. Food Agric.* **2019**, *99*, 1804–1811. [CrossRef] [PubMed]
13. García-Esparza, M.J.; Abrisqueta, I.; Escriche, I.; Intrigliolo, D.S.; Álvarez, I.; Lizama, V. Volatile compounds and phenolic composition of skins and seeds of '*Cabernet Sauvignon*' grapes under different deficit irrigation regimes. *J. Grapevine Res.* **2018**, *57*, 83–91.
14. Zhu, Y.; Taylor, C.; Sommer, K.; Wilkinson, K.; Wirthensohn, M. Influence of deficit irrigation strategies on fatty acid and tocopherol concentration of almond (*Prunus dulcis*). *Food Chem.* **2015**, *173*, 821–826. [CrossRef] [PubMed]
15. Meilgaard, M.C.; Civille, G.V.; Carr, B.T. *Sensory Evaluation Techniques*, 4th ed.; CRC Press: Boca Raton, FL, USA, 1987.
16. Wei, S.; Ang, T.; Jancenelle, V.E. Willingness to pay more for green products: The interplay of consumer characteristics and customer participation. *J. Retail. Consum. Serv.* **2018**, *45*, 230–238. [CrossRef]
17. Rajagopalan, P. A Study on Consumer's Perception and Purchase Intentions towards Eco-Friendly Products. *Asian J. Res. Soc. Sci. Humanit.* **2016**, *6*, 1794–1802. [CrossRef]
18. Noguera-Artiaga, L.; Lipan, L.; Vázquez-Araújo, L.; Barber, X.; Pérez-López, D.; Carbonell-Barrachina, Á.A. Opinion of Spanish Consumers on Hydrosustainable Pistachios. *J. Food Sci.* **2016**, *81*, S2559–S2565. [CrossRef]
19. McCutchan, H.; Shackel, K.A. Stem-water Potential as a Sensitive Indicator of Water Stress in Prune Trees (*Prunus domestica* L. cv. French). *J. Am. Soc. Hortic. Sci.* **1992**, *117*. [CrossRef]
20. Vázquez Araújo, L.; Chambers, D.; Carbonell-Barrachina, A. Development of a sensory lexicon and application by an industry trade panel for *turrón*, a European Protected Product. *J. Sens. Stud.* **2012**, *27*. [CrossRef]
21. Nix Sensor Ltd. Need a Quick and Free Color Converter? 2018. Available online: https://www.nixsensor.com/free-color-converter/ (accessed on 23 December 2018).
22. Cano-Lamadrid, M.; Lech, K.; Michalska, A.; Wasilewska, M.; Figiel, A.; Wojdyło, A.; Carbonell-Barrachina, A.A. Influence of osmotic dehydration pre-treatment and combined drying method on physico-chemical and sensory properties of pomegranate arils, cultivar Mollar de Elche. *Food Chem.* **2017**, *232*, 306–315. [CrossRef]
23. Contador, L.; Robles, B.; Shinya, P.; Medel, M.; Pinto, C.; Reginato, G.; Infante, R. Characterization of texture attributes of raw almond using a trained sensory panel. *Fruits* **2015**, *70*, 231–237. [CrossRef]
24. Crisolar, Píldoras de Conocimiento: Los Calibres en Las Almendras. 2018. Available online: http://blog.crisolar.es/?p=483 (accessed on 23 December 2018).

25. Vermeir, I.; Verbeke, W. Sustainable food consumption among young adults in Belgium: Theory of planned behaviour and the role of confidence and values. *Ecol. Econ.* **2008**, *64*, 542–553. [CrossRef]
26. Lawless, H.T.; Heymann, H. *Sensory Evaluation of Food: Principle and Practices*, 2nd ed.; Springer: New York, NY, USA, 2010; pp. 507–534.
27. Reche, J.; Hernández, F.; Almansa, M.S.; Carbonell-Barrachina, Á.A.; Legua, P.; Amorós, A. Effects of organic and conventional farming on the physicochemical and functional properties of jujube fruit. *LWT* **2019**, *99*, 438–444. [CrossRef]
28. Rinaldi, R.; Amodio, M.L.; Colelli, G.; Nanos, G.D.; Pliakoni, E. Effect of deficit irrigation on fruit and oil quality of 'Konservolea' olives. *Acta Hortic.* **2011**, *924*, 445–451. [CrossRef]
29. Pérez-Sarmiento, F.; Miras-Avalos, J.M.; Alcobendas, R.; Alarcón, J.J.; Mounzer, O.; Nicolás, E. Effects of regulated deficit irrigation on physiology, yield and fruit quality in apricot trees under Mediterranean conditions. *Span. J. Agric. Res.* **2016**, *14*, e1205. [CrossRef]
30. Sotiropoulos, T.; Kalfountzos, D., II; Aleksiou, I., II; Kotsopoulos, S., II; Koutinas, N. Response of a clingstone peach cultivar to regulated deficit irrigation. *Sci. Agric.* **2010**, *67*, 2. [CrossRef]
31. Doll, D. Impacts of Drought on Almond Production. *West. Fruit Grow.* **2014**, *134*, S8–S10.
32. Cano-Lamadrid, M.; Galindo, A.; Collado-González, J.; Rodríguez, P.; Cruz, Z.N.; Legua, P.; Burló, F.; Morales, D.; Carbonell-Barrachina, Á.A.; Hernández, F. Influence of deficit irrigation and crop load on the yield and fruit quality in Wonderful and Mollar de Elche pomegranates. *J. Sci. Food Agric.* **2018**, *98*, 3098–3108. [CrossRef]
33. Verdú, A.; Serrano-Megías, M.; Vázquez-Araújo, L.; Pérez-López, A.J.; Carbonell-Barrachina, A.A. Differences in Jijona *turrón* concepts between consumers and manufacturers. *J. Sci. Food Agric.* **2007**, *87*, 2106–2111. [CrossRef]
34. Conesa, M.R.; de la Rosa, J.M.; Artés-Hernández, F.; Dodd, I.C.; Domingo, R.; Pérez-Pastor, A. Long-term impact of deficit irrigation on the physical quality of berries in 'Crimson Seedless' table grapes. *J. Sci. Food. Agric.* **2015**, *95*, 2510–2520. [CrossRef]
35. Vallverdú, X.; Girona, J.; Echeverria, G.; Marsal, J.; Behboudian, M.H.; Lopez, G. Sensory Quality and Consumer Acceptance of 'Tardibelle' Peach are Improved by Deficit Irrigation applied during Stage II of Fruit Development. *HortScience* **2012**, *47*, 656–659. [CrossRef]
36. Anderson, J.R. Concepts of Food Sustainability. In *Encyclopedia of Food Security and Sustainability*; Ferranti, P., Berry, E.M., Anderson, J.R., Eds.; Elsevier: Oxford, UK, 2019; pp. 1–8.
37. Cano-Lamadrid, M.; Vázquez-Araújo, L.; Sánchez-Rodríguez, L.; Wodylo, A.; Carbonell-Barrachina, A.A. Consumers' opinion on dried pomegranate arils to determine the best processing conditions. *J. Food Sci.* **2018**, *83*, 3085–3091. [CrossRef] [PubMed]
38. Pagès, J.; Berthelo, S.; Brossier, M.; Gourret, D. Statistical penalty analysis. *Food Qual. Preference* **2014**, *32*, 16–23. [CrossRef]
39. Ministerio de Agricultura, Pesca y Alimentación. Anuario de Estadística Avance. 2019. Available online: https://www.mapa.gob.es/es/estadistica/temas/ (accessed on 11 February 2019).

© 2019 by the authors. Licensee MDPI, Basel, Switzerland. This article is an open access article distributed under the terms and conditions of the Creative Commons Attribution (CC BY) license (http://creativecommons.org/licenses/by/4.0/).

Article

What Is "Natural"? Consumer Responses to Selected Ingredients

Edgar Chambers V, Edgar Chambers IV * and Mauricio Castro

Center for Sensory Analysis and Consumer Behavior, Kansas State University, Manhattan, KS 66506, USA; echamber@ksu.edu (E.C.); mauriciod@ksu.edu (M.C.)
* Correspondence: eciv@ksu.edu; Tel.: +1-785-532-0166

Received: 22 March 2018; Accepted: 16 April 2018; Published: 23 April 2018

Abstract: Interest in "natural" food has grown enormously over the last decade. Because the United States government has not set a legal definition for the term "natural", customers have formed their own sensory perceptions and opinions on what constitutes natural. In this study, we examined 20 ingredients to determine what consumers consider to be natural. Using a national database, 630 consumers were sampled (50% male and 50% female) online, and the results were analyzed using percentages and chi-square tests. No ingredient was considered natural by more than 69% of respondents. We found evidence that familiarity may play a major role in consumers' determination of naturalness. We also found evidence that chemical sounding names and the age of the consumer have an effect on whether an ingredient and potentially a food is considered natural. Interestingly, a preference towards selecting GMO (genetically modified organisms) foods had no significant impact on perceptions of natural.

Keywords: natural; ingredient; consumer; perception

1. Introduction

Humans have an instinctual desire for natural products and processes [1]. Consequently, the natural food market has grown substantially in recent decades. The Nielsen Global Health and Wellness Survey [2] found that in addition to eating fewer fats and sugary sweets, 60% of those dieting in North America were eating more "natural" and "fresh" foods, while 56% of Europeans and 68% of Latin Americans said the same. Customers said that the three "most desirable attributes are foods that are fresh, natural and minimally processed." Chambers and Phan [3] showed that sensory aspects are the most important consideration in eating behavior, but health also is important in many food selections, which suggests that customers who say that "fresh" and "natural" are key considerations may be using those terms at least in part to refer to sensory and health properties. Other authors [4] have shown that using the term "natural" has an impact on liking or the hedonic sensory experience of products. Foods that are considered natural and not genetically modified had the most important attributes. The Neilson data [2] also found that labels that referenced ideas that consumers already had about the product were purchased more than other foods.

The appeal of naturalness is two-fold, with many people finding it healthier and more moral. Natural food is considered to be a more moral choice over artificial foods [5]. Bratanova et al. [6] found that moral food choices, through a sense of moral satisfaction, make food subjectively taste better, which serves as a reward mechanism for buying the product. Natural food also is considered to be healthier [5,7] with natural meat providing a mediational effect on the risk of colon cancer associated with meat consumption. Lay person comments on redefining natural submitted to the United States Food and Drug Adminisgtration (FDA) support the theory that natural is associated with healthful food [8].

The problem in reviewing naturalness is that what constitutes a natural food in scientific studies may be different among studies and may not be the same as the definition that consumers use [1]. Many studies have used the same criteria as "organic" food as their criteria for naturalness, while others have used the presence of some degree of natural ingredients to constitute naturalness. This variation in definitions causes a massive variation in results and muddies the waters of what exactly natural means for both data collection and marketing purposes for companies in the food business. The confusion is corroborated by consumer comments sent to the FDA [8] during the comment collection process, with the definitions given ranging from products that are "are healthy in some way" to anything that is "not fresh should not be considered healthy". Similarly, Dominick et al. [9] found that 80% of respondents thought that "all natural" products would mean no antibiotics, no hormones, and no preservatives added in processing, with 60% of respondents sayings that natural food would have improved animal welfare practices, improved nutritional value, and improved food safety. A study in the Netherlands [10] found that those people who scored high on the scale of "prefers natural products" appeared to be most interested in "prevention-oriented motivations" (i.e., motivations for health and not eating foods they thought were unhealthful).

Other problems are that studies focus only on a single aspect to determine naturalness (for example, foods in packages [11]), or only on a single aspect of naturalness (e.g., health, environment, additives, etc.). In addition, some authors [12] have shown that natural may not be as holistically equivalent as others have suggested. For example, in a study in France, the respondents differentiated that health was not equivalent to natural, instead it was divided into a "natural" component associated with chemicals or GMOs (genetically modified organisms) and a component more related to disease that was not associated with natural.

The present research aims to investigate what percentage of the population considers various ingredients as natural, without giving participants a definition of natural. Such a procedure was used to best reflect what consumers believed was "natural", without biasing them toward a particular point of view.

2. Materials and Methods

Qualtrics (Provo, UT, USA), an online survey company, was used to recruit 630 subjects nationwide, with proportions divided to represent the four demographic regions of the United States (U.S.) (Northeast, South, Midwest, and West), as defined by the U.S. Census Bureau. A gender quota was set (50% male and 50% female) for all the different age groups. The participants were divided into three age classes—(1) 18–34 years old; (2) 35–54 years old; and (3) more than 54 years old—and 100 respondents per demographic group (gender X age division). Additional information was obtained, such as education level and number of adults and children under the age of 18 in the household. The participants did not receive a financial incentive for completing the online survey, but the Qualtrics database has a reward system in order to compensate the respondents for their time and collaboration.

The questionnaire (see the sample in the Supplementary Material) was developed to evaluate consumer responses to whether a material was considered "natural" or not. A Check All That Apply format was used for the questionnaire, which simply required that the respondents check each ingredient they believed would qualify as "natural". A total of 20 ingredients (Table 1) found in human foods in the U.S. was selected to represent a range of ingredients that consumers might find as common or unusual, common vs chemical names, and various sources and forms. Questions related to food and health beliefs (e.g., caring about ingredients, selecting GMO products, trying to eat for enjoyment) were also included to help determine the impacts of various health or food beliefs. The survey was conducted in English using Qualtrics survey software and was approved by the Institutional Review Board of Kansas State University.

The data were analyzed in Excel (Microsoft Office Pro ver. 2013) using descriptive statistics and chi-square tests for significance, considering a p-value under 5% to be significant.

Table 1. Demographics Percentages *.

Demographic Characteristics	Variables	Percentage
Gender	Men	50%
	Women	50%
Age	18–34	33%
	35–54	33%
	≥55	33%
Level of Education	Primary School or Less	1%
	High School	45%
	College or University	53%
Number of Adults Living at Residence	1	27%
	2	48%
	3	15%
	4	7%
	5 or More	3%
Number of Children Living at Residence	0	67%
	1	16%
	2	10%
	3	4%
	4 or More	2%

* Percentages were based on 630 consumers in total.

3. Results

The results of the study found that corn, wheat flour, black beans, and soybeans were all considered natural by at least 60% of consumers (Table 2). Only those four foods, as well as sugar and salt, were perceived as natural by more than half of consumers. Xanthan gum, insect powder, maltodextrins, butylated hydroxyanisole (BHA), and sodium acid pyrophosphate (SAPP) were considered natural by the fewest consumers, with less than 10% of consumers considering those ingredients as natural.

Several inconsistencies were found in consumer opinions; most notable were the relationships between gluten and wheat flour, baking soda and sodium bicarbonate, and the differences between the various flours (pea, wheat, and sorghum). Wheat flour was considered natural by nearly three times as many people as gluten (a major protein in wheat flour). Baking soda also was more likely to be considered natural compared with sodium bicarbonate (the chemical name of baking soda). Interestingly, fewer than half as many people considered sorghum and pea flours natural as did wheat flour.

The data were cross tabulated between different demographic and belief factors asked during the survey. Age group showed the most significant ($p < 0.05$) differences; sorghum flour, black beans, lecithin, corn syrup, and molasses were all more likely to be considered natural by participants older than 55. Only insect powder and gluten were significantly ($p < 0.05$) less likely to be considered natural by those over 55 than other age groups. Gender was found to have a significant ($p < 0.05$) effect on three items—lecithin, corn syrup, and maltodextrins—with more men being more likely to consider each of those ingredients natural. Education was found to have no significant impact on perception of naturalness.

A number of demographic/attitudinal subgroups influenced just one or two ingredients. Those who responded that they agree with the statement "I care about the ingredients in my food" were significantly ($p < 0.05$) more likely to consider black beans and baking soda natural. People agreeing with the statement "I always follow a healthy diet" were more likely ($p < 0.05$) to consider both sorghum flour and sodium bicarbonate not natural. Respondents who agreed with the statement "I don't worry about naturalness" were significantly ($p < 0.05$) less likely than expected to consider black beans and molasses natural. Those who stated "I look for non-GMO products" were not significantly different in their responses for any ingredient from those who did not choose that statement.

Table 2. Percentage of people saying that an ingredient was natural.

Ingredients	Percent Saying Ingredient Was Natural
Corn	69%
Wheat Flour	63%
Black Beans	63%
Soybeans	62%
Salt	54%
Sugar	54%
Molasses	45%
Canola Oil	42%
Pea Flour	38%
Baking Soda	33%
Sorghum Flour	24%
Gluten	23%
Corn Syrup	20%
Lecithin	14%
Sodium Bicarbonate	13%
Xantham Gum	8%
Insect Powder	7%
Maltodextrins	5%
Butylated Hydroxyanisole (BHA)	4%
Sodium Acid Phyrophosphate (SAPP)	3%

4. Discussion

The main results showed that no ingredient was determined to be natural by all people. In fact, no ingredient listed in the survey was found to be natural by more than 70% of U.S. consumers. This suggests that consumers have multiple reasons for not believing that certain ingredients are natural and, clearly, differ in the importance they place on various characteristics. This concept of complexity in consumer expectations is supported by the U.S. Federal Trade Commission who decided not to define natural in 1983 because, in part, "consumer expectations were too complex to address" [13].

We found it especially surprising that that corn and soybeans were considered to be natural by over 60% of consumers even though 92% of corn and 94% of soybeans are genetically modified in the United States [14]. There was also no statistical difference between those responding that they look for non-GMO products versus those who do not usually look for non-GMO products. Both products are highly marketed, familiar to consumers, and are considered healthy foods in general. The discrepancy between the two products is likely caused by the perceived healthiness and familiarity bias. However, some consumers definitely believe that products from corn, such as high fructose corn syrup, and the use of soy in some "natural" products are not appropriate, as noted by lawsuits over such products [13].

We note that items with chemical names tend to be characterized by almost all consumers as not natural. This is true even for ingredients that are manufactured, as well as those that are derived, from natural sources (e.g., xanthan gum). In addition, some ingredients, such as insect powder, which consumers may believe to be an adulterant or disgusting rather than a wholesome ingredient, cannot be natural. Shim et al. [15] found safety perception and knowledge of naturalness increased by 60% with communication programs that help customers become more familiar with the product.

The discrepancies noted in the results for items such as wheat flour vs gluten also have likely reasons. Those data suggest that particular portions of an ingredient or having to further process an ingredient to extract fractions (wheat flour vs gluten) or the perceived dietary constraints of an ingredient name (gluten) negatively impacts the perception of natural. Similarly, the use of a familiar vs chemical name (baking soda vs sodium bicarbonate) impacts the perception of natural, with the chemical name greatly reducing the perception of naturalness. The difference in flour types suggests that familiarity (wheat flour, which is a common ingredient, vs sorghum or pea flours, which are less common) plays a role in defining "natural". Evans et al. [16] proposed the hypothesis that lesser known ingredients will be perceived to be less natural then well-known ingredients. Although these reasons may seem obvious or valid on the surface to both researchers and consumers, they also clearly show the difficulty in defining what is natural. If wheat flour is considered natural (63% of consumers said it was), but sorghum flour, which is more likely to be GMO-free and perhaps more likely to be organic, is not considered natural (only 23% of consumers thought so), what then constitutes reality? A clear definition could help in labeling, but getting to that point may pose huge problems because of the disconnect between science and the consumer's perception or knowledge [13].

We believe that the familiarity bias explains why men were more likely than women ($p < 0.05$) to respond that maltodextrins and lecithin were natural. Both products are frequently used in protein powders and protein snacks, which tend to be marketed and consumed more by men. The clearest evidence of the familiarity bias is the difference in those saying that baking soda was natural compared with those who said that sodium bicarbonate was natural. As shown in several studies, chemical names or "e numbers" (used in Europe on ingredient labels) [7] tend to be considered less natural by consumers.

Insect powder and pea flour appear to be good examples of food neophobia affecting the perceived naturalness of ingredients. An individual's expectations toward food products play a critical role in consumer psychology [17], and flours and powders are not what U.S. consumers' think of when they think of insects and peas. There may be a neophobic response to their acceptance, which will influence the perception of naturalness. It is likely that this effect is amplified by the fact that insect powder and pea flour are "novelty" foods, to which most Americans have had little exposure. It also is likely that the physical change (manufacturing) of turning the ingredient into a powder is driving down the perception of naturalness. Rozin [18] found that grinding peanuts into peanut butter reduced overall perceptions of naturalness by 8.3%.

The fact that those over 55 years of age were more likely to consider a product natural is interesting. As Walker et al. [19] found in 1991, the biggest risk factors for dietary inadequacies were loneliness and social isolation. However, in recent years, technology and culture changes in the United States have mitigated these issues, because those over 55 now have more access to the internet, social programming and healthcare services, and nutritional information. With increasing life expectancy, older adults are taking a bigger role in their own health. The only ingredient for which the results were lower than expected for that age group was gluten, and we expect that this was because of the gluten scare that has occurred over the last few years. This, directly refutes claims made by Rozin et al. [18] that age and other demographic factors do not contribute much to the perception of naturalness and suggests that changes in the market in the last 5–6 years may be changing the perceptions of some consumers.

We believe that lack of familiarity explains the low percentages for ingredients such as molasses and sorghum. The decline in the popularity of molasses in the U.S. after World War II and the fact that few human foods contain grain sorghum (in the past, it was used primarily for animal feed in the U.S.) means that both products are not widely known, and thus, people do not say that they are natural.

The results also indicate an ideological gap between "what people think constitutes a natural product" and "what products are natural". As discussed by Román et al. [1] studies about naturalness usually research what factors make a product natural and how product manipulation affects naturalness. Instead, our study focuses on the consumers' perception of a holistic ingredient. Thus, rather than examining ingredients in the context of whether certain aspects of an ingredient are natural, our study examined ingredients as a single concept. Thus, an ingredient may not be considered natural when considered against a particular criterion but could be considered natural by consumers simply have an image of a product (in this case an ingredient) in their minds outside of the context of a specific set of values or motivations.

One limitation of this study is that consumers were not asked to explain why they considered a food natural or not. The survey provides only an understanding of the likelihood that an ingredient is perceived as natural. Thus, we can speculate, but not cannot be sure of the reasoning of consumers for their responses. Further research is needed for such information.

5. Conclusions

Our results show that consumers do not agree on which ingredients are natural and which are not, with a few exceptions. Thus, it is likely that consumers have different definitions for what constitutes natural. This allows for greater consumer confusion and frustration when trying to find and purchase natural food products and ingredients. We believe that that further research is needed to help determine what consumers believe constitutes natural. Moreover, we think that there is a lack of research into what ingredients people consider natural, not just what makes a food natural. Further research in this area, including in other countries and to determine the reasons people believe various ingredients may not be natural, is necessary.

One key aspect that is lacking appears to be agreement in the scientific community and a coherent educational campaign as to "what is natural". Research already shows that disagreement exists at the product (food or beverage) level. If further research shows that even at the ingredient level consumers cannot agree on what constitutes natural in part because there is confusion over what the term means, then educational efforts to explain the term are needed. In addition, better consumer information about some ingredients (e.g., sorghum) or new products is needed in order to overcome what we believe is a lack of familiarity that can breed unease with consumers.

Supplementary Materials: The following are available online at http://www.mdpi.com/2304-8158/7/4/65/s1. Supplementary Material: Natural Portion of U.S. Survey.

Author Contributions: E.C.IV and M.C. conceived of and designed the work; E.C.V analyzed the data and wrote the article; E.C.IV, E.C.V, and M.C. revised and proofread the article.

Conflicts of Interest: The authors declare no conflict of interest.

References

1. Román, S.; Sánchez-Siles, L.M.; Siegrist, M. The importance of food naturalness for consumers: Results of a systematic review. *Trends Food Sci. Technol.* **2017**, *67*, 44–57. [CrossRef]
2. Nielsen. *We Are What We Eat Healthy Eating Trends around The World*; The Nielsen Corporation: New York, NY, USA, 2015.
3. Phan, U.X.T.; Chambers IV, E. Application of an eating motivation survey to study eating occasions. *J. Sens. Stud.* **2016**, *31*, 114–123. [CrossRef]
4. Apaolaza, V.; Hartmann, P.; López, C.; Barrutia, J.M.; Echebarria, C. Natural ingredients claim's halo effect on hedonic sensory experiences of perfumes. *Food Qual. Pref.* **2014**, *36*, 81–86. [CrossRef]
5. Rozin, P.; Spranca, M.; Krieger, Z.; Neuhaus, R.; Surillo, D.; Swerdlin, A.; Wood, W. Preference for natural: Instrumental and ideational/moral motivations, and the contrast between foods and medicines. *Appetite* **2004**, *43*, 147–154. [CrossRef] [PubMed]
6. Bratanova, B.; Vauclair, C.; Kervyn, N.; Schumann, S.; Wood, R.; Klein, O. Savouring morality. Moral satisfaction renders food of ethical origin subjectively tastier. *Appetite* **2015**, *91*, 137–149. [CrossRef] [PubMed]
7. Siegrist, M.; Sütterlin, B. Importance of perceived naturalness for acceptance of food additives and cultured meat. *Appetite* **2017**, *113*, 320–326. [CrossRef] [PubMed]
8. Museum of Food and Drink. Questionnaire with Multiple Responses. US Food and Drug Administration 2016. Available online: https://www.regulations.gov/document?D=FDA-2014-N-1207-7334 (accessed on 12 March 2018).
9. Dominick, S.R.; Fullerton, C.; Widmar, C.J.O.; Wang, H. Consumer Associations with the "All Natural" Food Label. *J. Food Prod. Mark.* **2017**, *23*, 1–14. [CrossRef]
10. De Boer, J.; Schösler, H. Food and value motivation: Linking consumer affinities to different types of food products. *Appetite* **2016**, *103*, 95–104. [CrossRef] [PubMed]

11. Binninger, A.-S. Perception of naturalness of food packaging and its role in consumer product evaluation. *J. Food Prod. Mark.* **2017**, *23*, 252–266. [CrossRef]
12. Sautron, V.; Péneau, S.; Camilleri, G.M.; Muller, L.; Ruffieux, B.; Hercberg, S.; Méjean, C. Validity of a questionnaire measuring motives for choosing foods including sustainable concerns. *Appetite* **2015**, *87*, 90–97. [CrossRef] [PubMed]
13. Parasidis, E.; Hooker, N.; Simons, D.T. Addressing consumer confusion surrounding "natural" food claims. *Am. J. Law Med.* **2015**, *41*, 357–373. [CrossRef] [PubMed]
14. U.S. Department of Agriculture Economic Research Service. Adoption of Genetically Engineered Crops in the U.S. USDA 2017. Available online: https://www.ers.usda.gov/data-products/adoption-of-genetically-engineered-crops-in-the-us.aspx (accessed on 22 March 2018).
15. Shim, S.M.; Seo, S.H.; Lee, Y.; Moon, G.I.; Kim, M.S.; Park, J.H. Consumers' knowledge and safety perceptions of food additives: Evaluation on the effectiveness of transmitting information on preservatives. *Food Control* **2011**, *22*, 1054–1060. [CrossRef]
16. Evans, G.; de Challemaison, B.; Cox, D.N. Consumers' ratings of the natural and unnatural qualities of foods. *Appetite* **2010**, *54*, 557–563. [CrossRef] [PubMed]
17. Tuorila, H.; Meiselman, H.L.; Bell, R.; Cardella, A.V.; Johnson, W. Role of sensory and cognitive information in the enhancement of certainty and liking for novel and familiar foods. *Appetite* **1994**, *23*, 231–246. [CrossRef] [PubMed]
18. Rozin, P.; Fischler, C.; Shields-Argelès, C. European and American perspectives on the meaning of natural. *Appetite* **2012**, *59*, 448–455. [CrossRef] [PubMed]
19. Walker, D.; Beauchene, R.E. The relationship of loneliness, social isolation, and physical health to dietary adequacy of independently living elderly. *J. Am. Diet. Assoc.* **1991**, *91*, 300–304. [PubMed]

 © 2018 by the authors. Licensee MDPI, Basel, Switzerland. This article is an open access article distributed under the terms and conditions of the Creative Commons Attribution (CC BY) license (http://creativecommons.org/licenses/by/4.0/).

MDPI
St. Alban-Anlage 66
4052 Basel
Switzerland
Tel. +41 61 683 77 34
Fax +41 61 302 89 18
www.mdpi.com

Foods Editorial Office
E-mail: foods@mdpi.com
www.mdpi.com/journal/foods